Shamanic Journeys Through Daghestan

First published by O Books, 2009
O Books is an imprint of John Hunt Publishing Ltd., The Bothy, Deershot Lodge, Park Lane, Ropley,
Hants, SO24 0BE, UK
office1@o-books.net
www.o-books.net

Distribution in:

UK and Europe
Orca Book Services
orders@orcabookservices.co.uk
Tel: 01202 665432 Fax: 01202 666219
Int. code (44)

USA and Canada
NBN
custserv@nbnbooks.com
Tel: 1 800 462 6420 Fax: 1 800 338 4550

Australia and New Zealand
Brumby Books
sales@brumbybooks.com.au
Tel: 61 3 9761 5535 Fax: 61 3 9761 7095

Far East (offices in Singapore, Thailand,
Hong Kong, Taiwan)
Pansing Distribution Pte Ltd
kemal@pansing.com
Tel: 65 6319 9939 Fax: 65 6462 5761

South Africa
Alternative Books
altbook@peterhyde.co.za
Tel: 021 555 4027 Fax: 021 447 1430

Text copyright Michael Berman 2008

Design: Stuart Davies

ISBN: 978 1 84694 225 9

A CIP catalogue record for this book is available
from the British Library.

Printed by Digital Book Print

Shamanic Journeys Through Daghestan

Michael Berman

BOOKS

Winchester, UK
Washington, USA

CONTENTS

Acknowledgements

If any copyright holders have been inadvertently overlooked, and for those copyright holders that all possible efforts have been made to contact but without success, the author will be pleased to make the necessary arrangements at the first opportunity.

The cover picture is a photo of an oil painting by Maka Bataishvili, an artist from the Republic of Georgia. For further details, please visit her website: www.maka.batiashvili.net Alternatively, if you are interested in buying any of her work and you are based in the UK, please visit www.caucasusarts.org.uk

Foreword

Sharing borders with Russia, Chechnya, Azerbaijan, and Georgia, it is of little surprise to see that Daghestan again finds itself caught in the maelstrom of change that has swept through the North Caucasus since the break up of the Soviet empire. Indeed, there are many who argue that Daghestan, along with its neighbours, has long been subject to the shifting sands of imperial rise and fall. From Greek and Roman, through European and Turkish, to Soviet conquest, the social-cultural composition of the area known today as Daghestan has changed dramatically, and continues to do so, relative to the ebb and flow of economic-political developments of a truly international character. Combined with this overarching context, of course, are the multilingual, poly-ethnic and plural religious identities which interact to generate a complex, and often fractious, set of internal dynamics. Fuelled by increasing international interest in its natural reserves of oil, gas, coal, and minerals, domestic tensions born of social-cultural difference continue to ignite into open conflict within and across the myriad peoples, tribes and clans which make up the heterogeneous country of Daghestan.

As it has ever been in the North Caucasus, external forces and internal processes combine to produce the thread by which the tapestry of life is woven by individual and community alike. As Michael Berman demonstrates so admirably in this book, oral traditions and folkloric histories have long served Daghestanis as a central means of rendering their ever-changing world meaningful. Subject to the combination of large-scale economic-political transformation and rapid social-cultural change, however, the modes of life which have long supported the celebration and transmission of these traditions and histories are progressively being undermined.

As social scientists across the disciplinary spectrum have

long since maintained, modern society is implacably de-traditionalising in its ongoing erosion of the social-cultural mechanisms by which traditional forms of meaning-making are performed in the present and passed down to subsequent generations. By gathering and reproducing *en bloc* the various stories contained in this book, Michael Berman does a valuable service in that he reminds us of the rich and variegated religious-cultural heritage of the Daghestani peoples. For rendering otherwise dispersed source materials readily accessible in a single volume, this book is to be congratulated.

The author, however, does much more than this. Michael Berman practises what he teaches and, as such, comes to the stories collected here with a wealth of practical knowledge from which the reader benefits greatly. At the same time, the author brings to bear a range of academic insight born of sustained reading of the relevant literature and reflection upon the most pertinent debates. Add to these ingredients the passion and commitment which have motivated the carrying to fruition of this project, and one is able to appreciate why the author's concept of 'shamanic story' is able to capture the subtleties and unlock the nuances of the materials reproduced in the following pages. I greatly enjoyed each of the stories contained in this book, just as I very much appreciated Michael Berman's interpretation of them. I trust you will too.

Andrew Dawson Lancaster University, UK

Chapter 1

Daghestan & the Shamanic Story

How to pin down the North Caucasus? You can't. You change worlds and centuries in an hour. In a single day you meet patriarchs and brigands, swaggering Mafiosi and shepherd boys; there are faces from ancient Greek pottery and strains of music from Turkey. None of it can or should be pinned down. The North Caucasus is a hall of a thousand distorting mirrors, each showing a different reflection, and people have long forgotten which are straight, which crooked (Smith, 2006, p.270).

Although the quote above comes from a book about Chechnya, it can equally as well be applied to Daghestan too. Known as the "land of the mountains,"[1] Daghestan lies immediately north of the Caucasus Mountains, and stretches for approximately 250 miles along the west shore of the Caspian Sea. It has been described as "the tip end of Europe. The Caucasus range is the boundary between the two continents … and the wall of separation between the Christian and the Mohammedan worlds" (Curtis, 1911, p.228). Today, however, the situation is of course not so clear cut.

With its mountainous terrain making travel and communication difficult, Daghestan is still largely tribal and, unlike in most other parts of Russia, the population (2,576,531 in 2002) is rapidly growing. Despite over a century of Tsarist control followed by seventy years of repressive Soviet rule, there are still 32 distinct ethnic groups, each with its own language, and Avar is the most widely spoken with about 700,000 speakers. With so many indigenous ethnic groups, Daghestan is unquestionably the most complex of the Caucasian republics.

In the lowlands can be found Turkic nomads: Kumyks, Noghays, and a few displaced Turkomans. In the northern highlands are the Avars, the Andis, Karatas, Chamalals, Bagwalals, Akhwakhs, Botlikhs, Godoberis, and Tindis. Still in the high valleys but going south toward the Georgian border are the Tsez (Dido), Ginukhs, Hunzibs, Khwarshis, and Bezhitas (Kapuchis). South of the Avar are the Laks, Dargwas, Kubachis, and Khaidaqs, all forming a related group of peoples. In one high village, standing apart from them, are the Archis, whose links lie further south with the so-called Lezgian peoples: the Aghuls, Tabasarans, and Rutuls. A few of the Lezgis and most of the Tsakhurs spill over into Azerbaijan in the south. Other Daghestanis who are restricted to northern Azerbaijan are the Kryz in one mountain village and three coastal ones, Budukhs (one mountain village), Udis (two mountain villages), and Khinalugs (one mountain village). There is also a group called "Mountain Jews" (Givrij or Dagchifut) who speak an Iranian language in Daghestan. They are sometimes called "Tats," but are not to be confused with the Muslim Tats further south on the Apsheron peninsula of Azerbaijan. In addition there are a few Daghestani Cossacks who are strongly assimilated to indigenous patterns.

Colarusso (1997) who compiled the above list, stresses that all thirty-two

> are distinct peoples, however small they may be, with their own languages, customs, costumes, arts, and architectures. Many are further subdivided by tribes, clans, and bloodlines. Conversely, most will traditionally form larger units for self-defence when threatened. This is particularly true of the smaller peoples of Daghestan. In ethnographic, social, and political terms the Caucasus is like a miniature continent.

To give some idea of the problems caused by the linguistic mix, despite the fact that Dargi and Avar are spoken by people living side by side with each other they are in fact mutually incomprehensible languages (see Chenciner et al, 1997, p.9). Multilingualism is therefore virtually universal. Nearly everyone speaks Russian in addition to their own language, and many have some command of several neighbouring languages too.

The Avars, incidentally, "are an offshoot of the people of that name who accompanied the Huns in their migration westward, and a small body of whom, bending their course to the south, crossed the Terek, and effected a settlement in the mountains of Daghestan, where their descendents still form a distinct tribe" (Ussher, 1865, p.144). In the eighteenth century, the Avars were the only group in the region to successfully defy the Persian invasion of Nadir Shah, and Shamyl, who fought a religious war so successfully for many years against the Russians in the last century, was also an Avar.

According to the 2002 Census, Northeast Caucasians (including Avars, Dargins and Lezgins) make up almost 75% of the population, Turkic peoples, Kumyks, Nogais and Azeris make up 20%, and Russians 5%. There are also forty or so tiny groups such as the Hinukh, numbering 200, or the Akhwakh, who are members of a complex family of indigenous Caucasians. Most of the ethnic groups "are subdivided into *tukhums*, or extended family clans, which traditionally did not intermarry and often fought long blood feuds. The *tukhum* managed the village affairs and laws. Today, the *tukhum* still functions as a unit, but to greatly varying degrees among different ethnic groups of the mountain land. [In addition,] … elders throughout the region play a vital role in ensuring the preservation of traditional rules within the family and, by extension, in society" (Smith, 2006, p.23). In the villages in Daghestan, the clans have their own tea houses in which their members gather.

3

As is the case throughout the Caucasus, it is the tradition for the eldest man to be master of the extended family unit in so far as he rules over disputes or regulates social relationships with the outside world. Within the confines of the home however, the eldest woman runs the show.

According to Chenciner, "Both in the USSR and Eastern Europe, when the Communist Party totalitarian system dissolved, there was little to take its place except national fascism and simplified religion" (Chenciner, 1997, p.28). Dramatic prose indeed, but really nothing more than a sweeping generalisation and what does "simplified" religion mean? For example, the forms of Orthodox Christianity that are now flourishing once again in neighbouring Georgia and in Armenia, and the exact difference between them, are surely far from simple.

In Daghestan 93% percent of the population is Muslim, consisting of Sunni Sufi orders that have been in place for centuries, with Christians accounting for much of the remaining 7%. There are also the "Mountain Jews".

To clear up a common misconception, it should be pointed out that the so-called Mountain Jews are not the Jews from ancient Babylon (who in fact survived until the 1950s in Baghdad), but the ones who came from Persia, where they had settled after the destruction of the Second Temple. These Jews arrived in Daghestan in two main waves of migration, much more recently, over the last 300 years in fact. And "There is no evidence to support earlier settlement, which has been confused with survivals of the Khazars, who converted to Judaism in the eighth century" (Chenciner, 1997, p.255).

For many years the Communist Party in Daghestan waged war against local culture, especially Islamic practices, through the atheistic Society of Godless Zealots. However, despite all their efforts, two Sufi groups are active in both Daghestan and neighbouring Chechnya: the Naqshbandis and the Qadiris. They practise the *zikr* chant, and this is how it has been described:

4

The hypnotic male chanting, swaying form a single voice to the roar of the whole company was like the sea. Only this was no normal sea: it may have begun there but soon it changed into the waters of the Old Testament Creation which *could* stir the forgotten roots of humanity, releasing unknown powers. The chant took everything along on its path or *tariqat* towards the mystic experience where man communicated directly with God ... After prolonged solemn chanting by the men standing in a stationary ring, they swayed slightly and juddered apparently uncoordinatedly from foot to foot and began their extraordinary movement. Each man stretched his left hand straight down, with the palm horizontal and open. He raised his right hand above his head and brought it down to clap against his left which he did not move. At the same time his right knee jerked up and stomped back on the ground and so the circular movement began, repeated again and again to a new loud rhythmic beat ... They seemed possessed. I was told that women also performed this *zikr* (Chenciner, 1997, p.212)

By "possessed", Chenciner probably means in a trancelike state, which is what shamans traditionally enter–by such means as listening to the sound of a drumbeat, chanting, or by using psychoactive drugs. And it would seem highly unlikely that a culture in which Sufi dancing is widely practised was not at one time a shamanic one.

What we find in Daghestan is very much an amalgam of practices. For example, on tombstones with Arab inscriptions, microcosmic maps can be found similar to those found in Kaitag embroidery. Animist and Islamic iconographies thus exist side by side (see Chenciner, 1997, p.92). And on the day of the new moon before Uraza Bayram at the end of Ramadan, it was customary for people to visit the graves of their ancestors, What this shows is that as in neighbouring Chechnya and Azerbaijan

5

too, "there has been an Islamicisation of traditional ancestor-worship" (Chenciner, 1997, p.95).

Even some mosques had pre-Islamic features,

> like the zoomorphic rafter terminals of dragon neck and head shape, rising over the arcade along the front long side of the old mosque in Djuli in Tabassaran`. This was similar to other finials on both sides of the surviving pagan shrine at Rekom, near Tsei in North Ossetia, where the sacrificed ram skulls were stacked on shelves against the log cabin walls. The same many-headed motif also appeared on tapestry carpets (Chenciner, 1997, pp.145-146).

Not only have animist practices been incorporated into Islamic ones, but they have also been combined with Soviet rituals, an example of this being the Lezgin Festival of Flowers, *Sukversovar*, which is held in Akhti every June.

> Local boys and girls climbed the mountain peak to gather magic flowers before dawn. They took food and wine and sang and danced all day. The best girl worker in the *kolkhoz* was crowned queen, and danced with a clown in a tall white sheepskin hat, differing from the older version, where a shah would wrestle and fight the clowns with his *kinjal* dagger. Finally, the queen and her attendants rode down the mountain with the flowers to give them to the old villagers for luck. Celebrations continued until midnight. But they now wear foreign costumes, because the festival must be international, and is organised by the Party *khudozhnik samodetni*, or self-action artist (Chenciner, 1997, p.248).

Like all the other nations in the Caucasus, and as the following quote shows, the Daghestanis have long been noted as being a hospitable people, and whoever breaks bread or eats salt with them is protected and defended with their lives. A Tartar farmhouse is a hotel for travellers–a free house of call for the homeless. No hungry man was ever sent from a Tartar threshold, the exercise of hospitality being the most important article in their creed (Curtis, 1911, p.229).

In fact, as Colarusso (1997) points out, so important are guests considered to be that

In theory the host would even give his life to defend a guest. In return, the guest is expected to act discretely and respectfully to his host and so bring honour to the host's family. Even a prisoner of war or an enemy could be treated hospitably if he had shown great valour. In the midst of a duel, it was possible for one adversary to seek a suspension of hostilities for a period of time as long as several days. The combatants would then resume their struggle at an agreed upon time and place.

And this is what Chenciner has to say on the way that the various peoples care for their guests:

Once, every family in Daghestan had a lamp in their window in case a stranger needing help passed by in the night. When a guest entered a mountain house, he was first offered the sweet water from their spring. Nowadays he was also poured a strong drink … If guests were from the host's village, he bade them goodbye at his gate. When they were from another village, he escorted them to the boundary of his village. If the guest was an important foreigner, he was accompanied to the frontier of the region, or even Daghestan, in a motorcade (Chenciner, 1997, pp.123-124).

As for living conditions, in the high reaches of Daghestan they can be extremely difficult due to the prolonged winters, coupled with the poor quality of the land. Men have to leave their families and go on long outings to earn money in the lowlands, and the women remain behind to maintain the households while they are away. To give some idea of how harsh the conditions can be, there is an Andi anecdote about a man who one day spread his burka on the ground for a midday nap and, on waking up, searched in vain for his plot of land until he found it hidden under his burka (see Hunt, 2007, pp.1-2).

Another reason life was hard were the natural barriers to travel, with transport always having been a problem in the mountains, carved up as they are by torrential rivers. Most roads are nothing more than bulldozed rocks or dirt, full of potholes (see Chenciner, 1997, p.130).

One advantage to living in the highlands though is that it ensures plenty of exercise, which probably helps to explain why so many people live to such ripe old ages.

Survival often depended on there being an adequate harvest, which was at the mercy of the elements, and ceremonies in which offerings were made to invoke the rain evolved from human sacrifice to save crops in times of drought.

In Karabudakent, after a prolonged drought, a meeting was announced from the minaret. Those with milk herds were asked to bring milk, and the rest to bring produce for the sacrificial meal. In a separate ceremony the women went up to their sacred place, while the men sacrificed a horned animal at the sacred well, dressed in skin coats turned inside out. They chanted a special prayer, shouting "Yes, we will have rain!" and then poured the water from the well over each other. Young men were made to jump into the cold water, after which they all went to the mosque for the meal (Chenciner, 1997, p.105).

Invocations for the sun were conducted along similar lines. "Girl mummers carried a doll-totem ... from house to house, singing songs of supplication to the sun, for which they received presents (Chenciner, 1997, p.106).

The Northeast Caucasian stone houses generally run on top of one another, the roof of the lower serving as the porch of the upper, as they cling to hill sides to form compact villages called *auls*. Traditionally the houses tended to be built together "on a south-facing amphitheatre shaped slope for defence, light and warmth" (Chenciner, 1997, p.135). Made of local stone, the houses generally have two floors: the first for livestock, the second for people. One of the reasons for this arrangement is that in winter it enables the people to live above the snow.

There was something else that could be found in every home:

Throughout the North Caucasus, every home possessed a wrought-iron hearth chain hanging down the chimney. Apart from the practical function of hooking up the cooking pot, the hearth chain has a central symbolic role for the family. ... The chain represented a microcosmic link between the fire and the food, or prosperity of the family, and the heavens above, where the family ancestors lived. It was another variation of the universal cosmic pillar separating heaven and earth (Chenciner, 1997, p.151).

The verb "separating" that Chenciner uses should perhaps be replaced here with "linking" or "connecting", as that was the traditional use that the Tree of Life, the Sacred Mountain, or Jacob's Ladder was put to, enabling the shaman or equivalent figure to travel between the different worlds in order to serve the community he or she represented, and to act as an intermediary on their behalf.

"Since the collapse of the Soviet Union in 1991, Daghestan is virtually the only part of the Caucasus that has not suffered inter-ethnic conflicts. This is probably because there are no dominant ethnic minorities due to the topology and linked ethnic mosaic of the region. In addition, that is likely to be the reason for the richness of their arts" (Chenciner et al, 1997, p.9). Since 2000 however, Daghestan has been the venue of a low-level guerilla war, bleeding over from Chechnya; the fighting has claimed the lives of hundreds of federal servicemen and officials–mostly members of local police forces–as well as many Daghestani national rebels and civilians.

There is a long history of conflict in the region prior to the Soviet era too, for Daghestan has "always been the prey of rival powers [and] ... has been plundered and ravaged in turn by all the Asiatic hordes" (Curtis, 1911, pp.228-229).

For millennia successive waves of conquerors have swept through the narrow coastal strip constricted at the 5,000-year-old city of Derbent. The road joins Russia and the Steppes to the north, Turkey to the west and Iran and the Middle East to the south. During the 6th century AD the Sassanian Persians built the double walls of Derbent up from the Caspian Sea to the mountains. After the Great Wall of China, the walls rank in size alongside Hadrian's Wall, or the triple walls of Constantinople (Chenciner et al, 1997, p.9).

It is said of the wall that it "extended from the present city of Derbent on the shore of the Caspian Sea, to the mountain of Koushan-Dagh near the western limits of Daghestan", and that "it was eighteen or twenty feet high and so thick that a squadron of cavalry could gallop along its top" (Curtis, 1911, p.233). And as for Derbent, it "was the conqueror's prize for Romans, Sarmatians, Sassanians, Georgians, Armenians, Khazars, Huns, Arabs, Mongols, Seljuk Turks, Ottoman Turks, Safavid Persians, Qajar Persians and Russians. It was not surprising that the ancient local tribes were great warriors. During the brief intervals of peace, they retreated to their independent mountain strongholds, to reunite again when faced with a fresh outside threat" (Chenciner et al, 1997, p.9).

As Colarusso (1997) points out, religious tolerance is a feature of the North Caucasus, where Sunni Muslims, Orthodox Christians, Jews, and pagans can all be found living amicably side by side, and this can probably be partly attributed to the fact that religion is socially and conceptually subordinated to ethnic identity throughout the region.

In the highlands there are mystical traditions of meditation and martial arts, which in Chechnya and Daghestan developed into Sufi practices. And in Daghestan holy men often have shrines, which are usually placed at the highest point of the village. Enigmatic relics of pagan beliefs persist throughout the Caucasus. In Daghestan, for example, there are many old beliefs surrounding animals, such as snakes, horses, and especially the bear. This last totemistic animal is associated with sacred rocks and even a half-bear half-man creature who features in the story *Bear's Ear*.

On the subject of the early beliefs of the people, prior to the adoption of Islam, we also know that the cult of tree veneration was widespread, and sacred trees and groves are to be found in Daghestan, Ossetia and Ingushetia even today

In Daghestan, the ancestors believed that trees or groves were occupied by spirits. In their folktales there are several instances, for example, when a comb is thrown or drops out over the hero's shoulder, it changes into a tree or a branch that shields him or her from the force of evil. Another manifestation of the cult is that, until today, in the mountains, it is considered an irreparable sin to cut down a tree. In some villages there it is also customary to spread out head-scarves or rags on trees that grow in cemeteries. If there are no trees there, a pole is planted among the tombstones and it is dressed up as a substitute tree. As an echo of ancient sacrifice to the tree spirits, it is usual to 'plant' a false tree on which is hung the carcass of a ram, in front of a shepherd's wattle shelter. The dried tree branch is called *karats* (Chenciner et al, 1997, p.34).

We also know that when magical healing was called for,

> [T]he Lezgin resorted to a witch doctor, who performed various rites. He collected earth from holy places or shrines, such as the Pyre of Suleiman on Mount Shalbuzdag, which was mixed with water and drunk as medicine. He also used the human-like Mary's hair, a fibrous plant which hangs from branches. If a prayer was said while a strand was tied round the wrist or ankle, magic powers of healing were activated (Chenciner, 1997, p.87).

In place of the term "witch doctor" that Chenciner uses, with its negative connotations, it might be better to substitute "shaman" or even "medicine man" here, as their main role was one of healing too.

On the other hand, as in other indigenous shamanic communities, it has to be said that rites were not only performed to bring about cures. "The Lezgins in particular had discovered a most effective way to kill: by burying some of the victim's hairs, knotted with a woollen string and a fatty sheep's tail in a sunny place. As the fat melted, so the victim would sicken and die on torment" (Chenciner, 1997, p.87).

In addition to the Lezgin healers, "In Daghestan Sufi sheikhs and some elder-women tattooers appear to have similar roles to shamans in North Asia and elsewhere" ((Chenciner et al, 1997, p.28). And because of their spiritual power, the North Caucasian Sufi sheikhs had a political role, as can be seen from the way in which they resisted the Tsarist Russians and then the

Communists during the 19th and 20th centuries. However, as in the neighbouring countries, the cult of the elders and ancestor worship play a more prominent role in the daily lives of the people, and the oral histories of shamans tend to be inconsistent individual stories rather than cultural models that have become generally adopted (see Chenciner et al, 1997, p.29).

***

There follows some notes on the themes that can be found in the stories.

As we might expect from a people who live what is, to a large extent, "a relatively communal life, but who suffered oppression in the past, either from outsiders or from their own petty princes, many of the tales are an attempt to 'hit back' at authority" (Hunt, 2007, p.4).

> Many of the tales, especially the humorous ones, also attack or poke fun at hypocrisy … or pomposity. Such attitudes may be associated with the fairly communal society of the mountain villages; such a community objecting to those pretending to be better than the rest (Hunt, 2007, pp.4-5).

According to Helma Everdina van den Berg (1965–2003), the specifically Dargi folktales from the region represent the eternal themes of the struggle of good and bad, the search for happiness and truth, the fight for justice etc. Among the Dargi tales, the anecdote is most widespread. The main characters of these stories always display the qualities of a people's hero. Among Daghestanian anecdotes, a regular theme concerns the exposure of the clergy. The people noticed in their direct encounters with the ministers of religion, that their words were not in accordance with their lifestyle. Features like hypocrisy, greed and treachery are exposed and ridiculed in various hilarious episodes … The anecdotes express a dislike of the clergy and present a satirical depiction of the feudal society at the same time" (Berg, 2001, p.10).

It is interesting to note that exposure of the clergy was also of interest to the Communists, and not only did this apply to the Muslim clergy but to the Jewish clergy too:

> One task that Communist ideologues assigned to literature throughout the Soviet Union was the campaign against religion. This ideological task is performed by Semyonov's lengthy tale *Oshnehoy en rabi Hesdil* (Rabbi Hesdil's Paramours, 1928), in which a Mountain Jewish rabbi is caricatured as a swindler, chronic drunkard, and incorrigible womanizer (Mikdash-Shamailov, 2002, pp.42-43).

The text that follows, which exemplifies this theme, was written down by Zapir Abdullaev and first published in Abdullaev and Gasanova (1959).

Why did you sit up there in the Sky?

The very first meeting of Malla Nasradin and the chief is usually spoken about in the following way.

They say the chief invited Malla Nasradin into his castle. Malla Nasradin arrived and when he looked, he saw that the house was filled with people. They apparently all sat on the floor in a circle. Only the chief apparently sat on a very high takhta (a takhta is a large and wide wooden bench placed in the room or on the verandah).

The mullah gave his regards to the chief, saying 'May you be healthy, God!'

'I am not God,' answered the chief, 'I (am) ...'

Malla Nasradin did not let him finish speaking: 'For you I am prepared to give my life, Izrael!' (Izrael is one of the four archangels).

'What are you saying?' said the chief. What kind of Izrael am I?'

'I don't know,' answered Malla Nasradin, 'when you are neither God, nor Izrael, why then are you not sitting down like the people, and why did you sit up there in the sky?'

***

Judging from the tale that follows, the theme would seem to cover exposure not only of the clergy, but of those in positions of power in other fields too (The text was written down in 1955 in the village of Deybuk, Dakhadaev district, by S. Gasanova and comes from the same volume referred to above).

The Ruler and the Wanderer

The Azay ruler met a wanderer.

The chief said: 'You have been to a lot of villages, you saw a lot of people, tell me a good story.'

'There is no better story on earth than about man's death,' said the wanderer.

'You told me a bad story,' said the chief in anger. 'Tell me another story,' said the ruler.

'The fact that, the man who dies does not return for a second time, is good too,' said the wanderer.

This story did not make the chief happy either.

'Leave my house, for you have only told me bad stories,' said the chief.

'Let me stay until I tell you another story,' said the wanderer. 'Hey ruler, if the people would not die, all the chiefs born before you would still be here. If they were present, there would be no reign for you. And if the dead would become live, they would rise up and would not leave you in your reign, they would destroy you.'

15

The chief became happy and let the wanderer stay in his house.

***

Dargi folklore is part of the multi-ethnic culture of Daghestan, together with the oral traditions of the Avar, Lak, Lezgi, Kumyk and other Daghestanian peoples. At the same time, it comprises the oral traditions of the various ethnic groups within Dargi (e.g. Tsudakhar, Akusha, Kubachi, Kaytag). Every Dargi subgroup created works in its own native dialect. The multi-dialectal situation of the Dargi people and the geographical isolation of the various subgroups lead to the creation of various distinctive local folklore traditions and therefore the folklore of each group has specific concepts and artistic forms, depending on the local geographical circumstances, daily life, culture and language (Berg, 2001, p.9).

On the other hand, this does not mean to say that all the stories are unique. The Avar tale *Bear's Ear*, for example, is a variant of the Chechen story *Golden Leaves*, which can be found in *The Shamanic Themes in Chechen Folktales*.

As for folktales that are specifically Jewish in origin, "Jewish folklore treats of Heaven and Earth, of Paradise and Hell, of Good and Evil, of the natural and the supernatural, of the spiritual and the material, of the sacred and the profane" Ausubel, 1948, p.xxi). In other words, Jewish folklore would initially seem to be no different to folklore from any other tradition. In some respects, however, it is perhaps different:

A large number of legends and myths, derived from their neighbors in Persia and Babylonia among whom the Jews lived for so many centuries after the Captivity, tell of angels and demons-all mediators between God and man's destiny of which he is the architect according to the good or evil of his conduct. In hundreds of other tales, with the humanizing intimacy of the true folklore spirit, there passes through a procession of the Patriarchs and the Prophets, of the Jewish kings and heroes, sages and scholars, saints and sinners, martyrs and renegades, rationalists and mystagogues, men of faith and also men of little faith. One of the objectives of all these tales is didactic-to hold up to the view of the Jew the inspiring example of his eminent forefathers in righteousness. They have still other objectives-to offer consolation and hope to the afflicted, to reconcile for the simple Jew the unhappy destiny of his people with his own trust in God, and also to explain to him those Scriptural passages and incidents that baffle his questioning mind (Ausubel, 1948, p.xxi).

In addition to the folktales, the folk arts and crafts from Daghestan are also worthy of note. The peoples are famous for their woven rugs, tapestries, and textiles, with each ethnic group having its own distinctive patterns. The Kubachis are renowned for their metal working, and their swords and daggers are among the finest examples of Islamic metal working in the world. Additionally, it is common for women to supplement the meagre incomes of their households with knitting. Socks, mittens, and sweaters are handmade by them at home, and they are then sold in Georgia or in Russia by the menfolk.

***

Let us now consider what is meant by the term "shamanic story" and why the tales chosen for analysis in this study can be said to exemplify the genre.

In her paper "South Siberian and Central Asian Hero Tales and Shamanistic Rituals", the Leipzig researcher Erika Taube suggests that

Folktales–being expressions of early stages of the development of human society–reflect reality: material culture, social relations, customs, [and] religious beliefs. When folktales were being formed and appeared as vivid forms of spiritual and artistic expression in correspondence with the general social development, those elements, which nowadays are usually regarded as fantastic creations of human mind, were strictly believed phenomena, i.e. they were accepted as facts. Therefore, it is not at all a new idea that such tales sometimes reflect shamanistic beliefs and conceptions (Taube, 1984, p. 344).

If they were forms of "artistic expression", however, then they could well have been regarded as such by those they were told to and we actually have no way of knowing whether they were "accepted as facts" or not. On the other hand, what we can show is that they do reflect shamanistic beliefs and conceptions, and this becomes apparent once we start to analyze them.

Sir James Frazer made a similar claim in his abridged version of *The Golden Bough*, first published in 1922: "folk-tales are a faithful reflection of the world as it appeared to the primitive mind; and that we may be sure that any idea which commonly occurs in them, however absurd it may seem to us, must once have been an ordinary article of belief" (Frazer, 1993, p.668). In reality, however, there is no way we can be certain that any idea that appears in such tales must once have been an ordinary article of belief as, not being able to get inside other people's minds, we cannot possibly know what was actually the case.

On the other hand, as Emily Lyle (2007) points out in the abstract to her paper "Narrative Form and the Structure of Myth", what we can be reasonably sure of is that "At each stage in transmission of a tale from generation to generation, modifications take place but something remains. Thus there is a potential for material to be retained from a time in the distant past when the narrative was embedded in a total oral worldview or cosmology." In view of the fact that in the past shamanism was widely practised in the region where the tale presented here originates from, it should therefore come as no surprise that a shamanic worldview and shamanic cosmology is to be found embedded in it.

Stories have traditionally been classified as epics, myths, sagas, legends, folk tales, fairy tales, parables or fables. However, the definitions of the terms have a tendency to overlap (see Berman, 2006, p.150-152) making it difficult to classify and categorize material. Another problem with the traditional terminology is that the genre system formed on the basis of European folklore cannot be fully applied universally.

Consider, for example, Eliade's definition of myth. For Eliade the characteristics of myth, as experienced by archaic societies, are that it constitutes the absolutely true and sacred History of the acts of the Supernaturals, which is always related to a "creation", which leads to a knowledge, experienced ritually, of the origin of things and thus the ability to control them, and which is "lived" in the sense that one is profoundly affected by the power of the events it recreates (see Eliade, 1964, pp.18-19). However, many stories are "lived" in the sense that one is profoundly affected by the events they recreate without them necessarily being myths. Moreover, many shamanic stories could be regarded as having the above characteristics but would still not necessarily be classified as myths.

Another problem encountered is that a number of the definitions of what a myth is are so general in nature that they tend to be of little value. For example, the suggestion that a myth is "a story about something significant [that] ... can take place in the past ... or in the present, or in the future" (Segal, 2004, p.5) really does not help us at all as this could be applied to more or less every type of tale.

Mary Beard, considering the significance of distinctions

between such categories as "myth," "legend," and "folk-tale," concludes that in fact no technical definition distinguishing these is wholly plausible, since matters of technical definition are not really the issue. "For these are value judgments masquerading as professional jargon; they are justifications of neglect–the dustbin categories for all kinds of mythic thinking that we would rather not treat as 'myth'" (see Winterbourne, 2007, p.15). Be this as it may, it is surely indisputable that we need some form of labelling for the categories in order to be able to refer to them, and the argument being presented in this study is that the time has come to revise them.

For this reason a case was argued in Berman (2006) for the introduction of a new genre, termed the shamanic story. This can be defined as a story that has either been based on or inspired by a shamanic journey, or one that contains a number of the elements typical of such a journey. Like other genres, it has "its own style, goals, entelechy, rhetoric, developmental pattern, and characteristic roles" (Turner, 1985, p.187), and like other genres it can be seen to differ to a certain extent from culture to culture. It should perhaps be noted at this point, however, that there are both etic and emic ways of regarding narrative (see Turner, 1982, p.65) and the term "shamanic story" clearly presents an outside view. It should also be pointed out that what is being offered here is a polytheistic definition of what the shamanic story is, in which a pool of characteristics can apply, but need not.

What these stories often show is that by carefully sifting through them for, in effect, red herrings such as material clearly added at a later date, and by carefully using external controlling factors such as archaeological, historical, and linguistic infor-mation, it is possible to a certain extent to reconstruct ancient beliefs from very remote periods, and that much more of the unwritten past may now be recoverable by such techniques than we ever realised before

Characteristics typical of the genre include the way in which the stories all tend to contain embedded texts (often the account of the shamanic journey itself), how the number of actors is clearly limited as one would expect in subjective accounts of what can be regarded as inner journeys, and how the stories tend to be used for healing purposes.

The healing effect of such stories derives from their dramatic potential to induce a psychological effect–the way in which they can free the reader from a debilitating self-image by focussing on his / her consciousness instead on a world of supernatural power. Additionally, through the use of narrative, shamans are able to provide their patients "with a language, by means of which unexpressed, and otherwise inexpressible, psychic states can be expressed" (Lévi-Strauss, 1968, p.198).

In his Foreword to *Tales of the Sacred and the Supernatural*, Eliade admits to repeatedly taking up "the themes of sortie du temps, or temporal dislocation, and of the alteration or the trans-mutation of space" (Eliade, 1981, p.10)[22], and these are themes that appear over and over again in shamanic stories too.

They are also frequently examples of what Jürgen Kremer, transpersonal psychologist and spiritual practitioner, called "tales of power" after one of Carlos Castaneda's novels. He defines such texts as 'conscious verbal constructions based on numinous experiences in non-ordinary reality, "which guide individuals and help them to integrate the spiritual, mythical, or archetypal aspects of their internal and external experience in unique, meaningful, and fulfilling ways" (Kremer, 1988, p.192). In other words, they are teaching tales.

As for the style of storytelling most frequently employed in shamanic stories, it can perhaps best be described as a form of magic realism, in which although "the point of departure is 'realistic' (recognizable events in chronological succession, everyday atmosphere, verisimilitude, characters with more or less predictable psychological reactions), ... soon strange discon-tinuities or gaps appear in the 'normal,' true-to-life texture of the narrative" (Calinescu, 1978, p.386).

Magic realism, a fusion of logic and nonsense, "can be seen as a foretaste of the changing worlds conditions, in which our minds, having mastered 'black and white'-or normal earthbound logic, must now begin to operate beyond the threshold of purely sense-based thinking to master a different, higher logic" (Hallam, 2002, p.48).

Though it might be better to replace the term "sense-based thinking" with "logical thinking", and though it is debatable or not whether it is necessarily "higher", what cannot be denied is that it is certainly a different way of operating, which is of course a large part of its attraction. For it gives us new options to work with.

Before concluding this introduction, it is worth making two further observations on the nature of the tales included in this study. "Since shamanism is so widespread, it is self-evident that the tales about shamans [and shamanic stories as defined in this volume] will be coloured by the narrative traits and modes of cultural expression specific to the various regions" (Hultkrantz, 1993, p.41).

Additionally, it should be noted that "In areas where shamanism has long been a thing of the past [as in Chechnya], many tales contain only vague, piecemeal or inaccurate recollections of shamans and their like" (Hultkrantz, 1993, p.51). On the other hand, however vague, piecemeal of inaccurate these recollections might be, this does not invalidate the classification of the tales they are embedded in as shamanic stories for the purposes of this study, or their value. In fact, just the reverse is the case, as in the absence of more concrete evidence, they may well help us to reconstruct how things used to be in a time when shamanism was undoubtedly prevalent in the region.

According to the finest traditions of storytelling, there are only a set number of archetypal story lines to draw upon and all of them relate to the human condition-to the struggle of becoming fully human. The heroes, or heroines, find themselves facing impossible conditions that can only be overcome by an initiation journey, during which the outer struggle is seen as symbolic of an inner struggle, and when this is won the outer dangers fall away (Hallam, 2002, p.47).

For the shaman, however, the outer struggle is perceived to be as real as the inner struggle, simply another form of reality, and the shamanic story is an account of just such a journey.

Chapter 2

The Sea Horse

There lived or did not live a *pachakh* [an oriental king], and he had three sons. Every day they used to visit their father, ask him about his health, listen to his counsel and wait for orders.

One day they saw their father was gloomy, like a storm-cloud, and cheerless, like a lonely cloud.

"What's wrong with you, father!" exclaimed his sons. "What are you so sad about?"

"My children!" answered the *pachakh*. "An amazing dream has plunged me into sorrow. I dreamed of a white sea-horse of rare beauty. When the first rays of the morning sun appeared over the sea, the horse leaped out on to the shore, galloped round the entire world three times and galloped away into the depths of the sea. With him, into the abyss, my heart fell too. That is the cause of my sadness".

"We swear to you father, that either we will die or we will obtain that horse for you!" said the sons. They saddled the best horses and got going on their way.

On the third day they arrived at a crossroads, at which stood a huge stone with the inscription: Ride to the right, and you will meet no danger. Ride to the left, and you will meet no danger. Ride straight on, and you will find death or good fortune".

The eldest brother decided to ride to the right, the middle one to the left, but the youngest one straight on. Both of the eldest brothers began trying to persuade the youngest one not to take the risk, but the *djigit* [a daring Caucasian horseman, who can perform all kinds of tricks on horseback] answered, "Fortune is like the tail of a cock on a windy day. Either I will perish or I will come back with success. You stay alive, and tell father of my fate". They parted, and each went his own way.

The youngest brother was riding for a long time. By night he was riding, and by day he was riding, until he rode into a dense forest. For a long time he was wandering in this forest, and he could not even see the end of it. For four months the *djigit* did not meet a single living soul.

At last he saw a large footprint; it was three *lokots* long, one *lokot* wide and a half *lokot* deep [a measure, now no longer in use, equal to one cubit or the length of the forearm-about 50cm]. The *djigit* rode along the tracks and saw a field in which stood seven vast castles. Around each castle was a fence of steel stakes and iron wire, and on each stake a human head. The *djigit* called into one castle and saw a huge *khart* [a character in Daghestan tales; an ogre, often female]. Without delaying to think, the young man rushed towards the giantess and began to suck her breast.

"Now you have become my son, and I your mother [According to ancient Daghestan custom, a man, after touching the breast of a woman with his teeth, becomes her adopted son]. Otherwise I would have dealt with you like this", muttered the *khart*. Then and there she seized the cat, fried it over the fire and swallowed it down. "And now tell me who you are and what you have come here for?"

The *djigit* told his adopted mother everything and asked her advice.

"I have seven sons, seven Narts", said the *khart*. "Soon they will come back from hunting. One or two of them will certainly know how to help you. But for now, hide in this chest, because my sons cannot endure even the scent of a man".

The *djigit* had not even had time to hide himself before there sounded a loud roar and a loud laugh: it was the *khart's* sons, the seven Narts coming. Each one of them had on his shoulder a plane tree with a deer's carcase fastened to it. The Narts began shouting, "Ey, mother, put on the pot!"

Suddenly they started to notice a smell and they shouted,

"There is the scent of a man here!"

"You have gone crazy, how could there be a man here! Probably you sensed him somewhere in the forest, stupid!" the giantess called her sons names and set about the cooking.

After feeding her sons plenty of meat and giving them plenty of *buza* [a fermented drink made from millet, buckwheat or barley], the *khart* asked them, "Which one of you knows about the white sea horse, which goes round the whole world in the blink of an eye?"

Six of them kept silent, and only the youngest answered, "I know such a horse. He belongs to the sea *pachakh*. At sunrise he comes out of the depths of the sea and flies three times around the world. After that he bathes in a small lake on the seashore, rolls in the sand, and then vanishes into the sea. Next to the lake grows a plane tree. On it hangs his golden saddle with a silver bridle".

When the Narts had gone to sleep, the *khart* released her adopted son from the chest, showed him the way to the lake and wished him success. After reaching the sea, the courageous *djigit* dug a hole, climbed into it and began to wait.

At dawn the handsome white horse leaped out of the sea and flew round the earth three times. When he started to roll in the sand, the youngest son jumped on him and, like a snake, clung to his neck.

Three times the horse jumped up to the heavens, and three times he descended to the earth, but he was simply unable to throw off the *djigit*. "You have tamed me, and from now on you are my master. Saddle me and tell me where to take you", said the human voice.

"Take me home, to father!" ordered his rider, and the horse started tearing along with the speed of the wind.

Thus they were riding until it got dark. Suddenly there flared up such a brilliant light that it became as light as day. The *djigit* climbed off his horse and saw that it was the feather of an

26

unknown bird shining. "Should I take it or leave it?" said he turning to the horse.

"If you take it you will regret it, and if you leave it you will regret it", answered the horse.

"In that case, I will take it!" exclaimed the young man. He hid the feather in his cap and rode on further. After some time they reached the walls of a town and made a halt. The horse began to nibble the grass, while the *djigit* wrapped himself up in his *burka* [a traditional Caucasian black cape of felt or goat's or sheep's wool] and went to sleep.

The town's inhabitants suddenly noticed that in the middle of the night it began to get as light as day, and they were very frightened. The khan of the town became even more frightened. He summoned a hundred of his soldiers and ordered them to find out what it was all about. When they came out beyond the town gates, they saw that the light was coming out of the sleeping man's cap. They shook the *djigit* and brought him to the khan.

"Who are you and where is your journey taking you?" asked the khan.

"My family is unknown, and I am simply wandering through the world".

"Won't you tell me why it suddenly has become light?" asked the khan.

"There is nothing simpler", answered the *djigit,* and he showed him the extraordinary feather.

The khan very much wanted to get hold of the bird that had such feathers, and he ordered menacingly, "I swear by the milk of my mother [This oath on the milk of the mother is considered sacred in Daghestan], that you will obtain that bird for me, otherwise I will order your head to be cut off!"

The *djigit* was obliged to give his oath and he set out after the bird. The white horse saw his master and asked, "What are you sad about, my friend?"

The *djigit* told him about the oath that the khan had made him give.

"Do not be upset" answered the white horse. "It is not a difficult matter. We will set off to that lake where I used to bathe. Every day three doves come flying there. They are the daughters of the sea king. They throw off their feathers and go in to bathe. Take the feathers of the youngest one and do not give them back, in spite of the beauty's urging. After that she will follow you".

As soon as they reached the lake the young man hid himself in the bushes and began to wait. At noon the three doves came flying in, turned into young girls and began to bathe.

The *djigit* did everything the white horse had told him to do and, regardless of her entreaties and requests, he did not return her plumage to the girl. Towards evening, when the daughters of the sea king were getting ready to fly away, the youngest said, "My sisters, farewell and be happy. I am obliged to stay here. Bring me my little casket".

Soon one dove brought a small cedar casket in its beak and quickly flew away again. The youngest sister dressed and went out on the shore. The *djigit* sat her behind his back and they galloped off.

"Tell me, *djigit*, who are you and where are you taking me?" asked the girl.

The young man told her everything and ended with these words: "Look: there is the town where the khan, whom you are supposed to marry, rules".

"Why give me away to him? It would be better if you married me yourself", asked the girl. The *djigit* kept silent, and they soon came riding to the palace.

The khan saw the girl; he started to tremble, his teeth began to chatter, and his eyes grew to the size of a fist. "I am going to marry her this very day. Get ready for the wedding!" shouted the khan.

"I am not going to marry such an old man", said the girl

turning obstinate. "If you want to get married, you must recover your youth".

"But how can that be done? After all, nobody can help me in that", whimpered the old khan.

"Order a large hole to be dug, fill it up with the milk of red cows and bathe in it. Then you will become young", advised the girl.

"In my realm there are not that many cows with a red colour", said the khan, becoming distressed.

"Send somebody to that hillock, let him wave this handkerchief, and the red cows will appear at once".

The khan made the townspeople dig the hole, and he sent his *nukyers* [originally the name of a member of the Mongol Khan's armed force, which later came to mean a servant or underling (often armed) of the khans and beks] to the hillock. One *nukyer* waved the handkerchief, and a large herd of red cows came running immediately. The hole was filled with milk, and the girl said, "Jump in, khan!"

But the khan was afraid, so then the girl told them to bring the very oldest people in the town. A hundred-year-old blind and deaf old man and a similar old woman appeared before her. The girl pushed them into the hole and then pulled them out again, young.

The khan at once jumped into the hole and began to sink. They say that even today he has not yet reached the bottom.

Meanwhile the *djigit* picked up the girl and galloped on further.

Along the road he landed in a certain town, and at the market he saw his eldest brother, who was serving in a bread shop. He was pale and dressed in pitiful rags. The youngest brother paid to redeem his brother from his master, and they all set off together for the middle brother.

The middle brother was a servant at a butcher's in another town. The youngest brother redeemed his middle brother too, and they all set out for home.

But envy was gnawing at the elder brothers, and they decided to kill the intrepid *djigit*. At a halt, when the youngest brother went to sleep, they threw him into a deep hole. After that they tried to catch his horse, but no way could they catch him. Then they seized the brother's bride, and shut the girl up in a tower. After that, the treacherous brothers presented themselves to the *pachakh*, their father. "O father! We have toured the entire world searching for the sea horse, but neither beneath heaven nor on the earth have we found such a horse".

"I do not need any kind of horse. Tell me what you have done with my youngest son?" asked the father, not satisfied.

"He went off along the dangerous road in spite of our persuasion to the contrary", answered the brothers, and they began to swear that they knew nothing more of their brother. The *pachakh* became terribly upset and declared mourning.

Meanwhile the brothers convinced their prisoner of the death of her bridegroom, and were trying to persuade her to marry one of them. But the girl invariably answered, "I will not make a pair with anyone who has served as a small-time salesman in a shop".

One day, from her window, the girl saw the white horse, who was ambling round the tower. She gave him a sign to come nearer. The horse came running to beneath the window. "Where is my bridegroom?" she asked, and found out that her bridegroom was still continuing to languish in the hole.

The girl began to weep, but the horse said, "Do not be sad! I will try to rescue him. Only get me a long rope" The girl cut off her tresses and twisted a long and firm rope from them.

"And now make a loop and throw it over my neck", asked the horse. The girl did just that.

Meanwhile the *djigit* had already despaired of getting out of the hole, when the strong rope fell at his feet, and up above he saw his faithful horse.

The courageous *djigit* rushed home like a whirlwind. His

brothers saw him and took to their heels: one to the West and the other to the East. It is said that they are running yet.

Back in the *pachakh's* palace, they arranged a splendid wedding, which went on for a very long time. At last it ended, and with it our tale.

***

The Sea Horse is an Avar tale and the Avar people form one of the main linguistic groups of the Lezg people of Daghestan.

> The mountainous nature of the country and the bravery of their men has protected the Avars from external pressure except for the most persistent invaders. In the eighteenth century, the Avars were the only group in the area to successfully defy the Persian invasion of Nadir Shah. Shamyl, who fought a religious war so successfully for many years against the Russians in the last century, was also an Avar; as were also many of the greatest men of Daghestan. The Avars are generally devout Moslems … and many of them make the pilgrimage to Mecca (Hunt, 2007, p.1).

However, this was clearly not always the case as the elements to be found in their folktales show.

The first point to note about the story is the way it starts. In place of *Once upon a time*, a convention that is also found in folktales from Georgia is employed–"There lived or did not live a *pachakh* [an oriental king], and he had three sons". This is how a variation on this typical introduction is explained in the story *The Tale of Tales*, which comes from *Yes and No Stories: A Book of Georgian Folk Tales* written by George Papashvily, an author who immigrated to the United States from Georgia in the 1920s: "[S]tories always begin the same way–There was, there was, and yet there was not–. It means that what comes after is true and true but then again not so true. Or perhaps it means that what is true for two men is not true for three."

What we have in this formulaic introduction is also an example of what has been referred to as "the alteration or the transmutation of space" (Eliade, 1981, p.10). As was pointed out in the opening Chapter, this is a theme that appears over and over again in shamanic stories, wherever they happen to come from.

Again and again in stories "we see how things appear in threes: how things have to happen three times, how the hero is given three wishes; how Cinderella goes to the ball three times; how the hero or the heroine is the third of three children" (Booker, 2004, p.229). Pythagoras called three the perfect number in that it represented the beginning, the middle and the end, and he thus regarded it as a symbol of Deity. And the number three plays an important part in our tale too. We learn that in the father's dream "the handsome white horse leaped out of the sea and flew round the earth three times", that "on the third day" the three sons arrived at a crossroads, and that "three doves came flying in, [and] turned into young girls".

This repetition of the number must surely be more than just coincidental. In fact, what we have in the tale is three times three, a trinity of trinities. The Pythagoreans believed that man is a full chord, or eight notes, and deity comes next. Three is the perfect trinity and represents perfect unity, twice three is the perfect dual, and three times three is the perfect plural, which explains why nine was considered to be a mystical number. Our tale certainly has a mystical element to it, and its connection to such symbolism clearly gives it greater significance. However, to suggest that the comparison with a trinity of trinities was intentional on the part of its author is perhaps, though interesting, a bit too far-fetched. One of the problems when it comes to considering symbolism is symbolic meaning can be read into almost anything and there is often no way of checking the interpretation.

As to the *pachakh*'s ailment, the cause of the sadness which he likens to his heart having fallen into an abyss, in shamanic terms it would appear to be a case of soul loss. Soul loss is the term used to describe the way parts of the psyche become detached when we are faced with traumatic situations. In psychological terms, it is known as dissociation and it often works as a defence mechanism, a means of displacing unpleasant feelings, impulses or thoughts into the unconscious. In shamanic terms, these split off parts can be found in non-ordinary reality and are only accessible to those familiar with its topography (see Gagan, 1998, p.9).

One of the techniques used as a form of therapy by both indigenous and neo-shamanic practitioners alike is what is known as soul retrieval. Soul in this context can be characterized as being our vital essence, where the emotions, feelings or sentiments are situated. The aim of soul retrieval is to recover the part of the client's soul that has been lost as this causes an "opening" through which illness can enter. The cause of this loss is believed to be due to an emotional or physical trauma that the client has been through, and it is the shaman's role to track down the lost soul part in non-ordinary reality and then to return it to the body (see Ingerman, 1993, p.23). Soul retrieval entails the shaman journeying to find the missing parts and then returning them to the client seeking help. In this particular tale, it is the youngest son who takes on this job.

The Caucasus is rich in folklore and this includes the Nart sagas, dramatic tales of a race of ancient heroes in which the figure of the all-wise Lady Satanaya is pivotal. She is mother to all the Narts, a fertility figure who is also an authority over her children and can be compared to the Greek Demeter. In *The Sea Horse* it is the *khart* who plays this role. The Nart giants originally featured in the Ossetic epic poems, and then spread south to Daghestan as characters in folktales and anecdotes.

In 'The Sea Horse', the hero avoids being eaten by a female ogre by "rushing towards the giantess and sucking at her breast". Her response is "Now you have become my son, and I your mother". Ritual sucking at the breast of the head female of a family is a well-known procedure in the Caucasus for the adoption of a person into the family [and certainly pre-dates the conversion of the people to Islam]. Her natural sons thereby become the foster brothers of the newly-adopted son (Hunt, 2007, p.5).

As Colarusso (1997) points out, this was also one of the traditional ways of putting an end to a blood feud: The killer "would sneak up on a woman from the victim's clan, seize her, tear open her blouse, and place his lips to her breast. Such an act, however fleeting, would suffice to create a kinship bond between the two clans which precluded further bloodshed".

In the last century, when a son was born to the ruler of the Karakaitags, he was sent from village to village to be suckled by all the women who could, in order to make him foster-brother of his entire generation. This was often a stronger tie than blood; Steder, a traveller-scholar of German origin, writing in 1797 noted that an Ingush murderer suckled his victim's mother at knifepoint, and so became part of her family, to avoid death by the blood-feud. Not long later, the Russian Karginov heard of an Ossete adulterer, who had been forced to kiss the breast of his beloved by her family, to terminate their relationship by making it incestuous (Chenciner, 1997, p.81).

Of the sea horse that belongs to the sea *pachakh*, we learn "that at sunrise he comes out of the depths of the sea and flies three times around the world. After that he bathes in a small lake on the seashore, rolls in the sand, and then vanishes into the sea".

The Waters have been described as the reservoir of all the potentialities of existence because they not only precede every form but they also serve to sustain every creation. Immersion is equivalent to dissolution of form, in other words death, whereas emergence repeats the cosmogenic act of formal manifestation, in other words re-birth (see Eliade, 1991, p.151).

"This same tale also includes the capture of a bride by the hero; in this case by stealing her feathered plumage while she is bathing" (Hunt, 2007, p.5). This reminds one of the numerous tales to be found in various traditions in which shape-shifting takes place. Animal skins are discarded to reveal human beings, who are then unable to return to their original forms due to their skins being stolen or hidden (see *The Robe of Feathers* in Berman, 2008a, for an example of this). And the ability to shape-shift is of course one of the attributes traditionally associated with shamans.

> Towards the end of the last century marriage by capture was still occasionally practised … but the groom still had to pay the kalim, or bride-price, otherwise a blood feud would ensue …The permitted degree of chastity varied from place to place, even amongst the Avars … Generally whilst the settlement of disputes was governed by a traditional type of 'common law', they could also be partly smoothed over by the decisions of the kadi, who based his decisions on the religious customary code, or the 'adats' (Hunt, 2007, p.3).

"The human-horse relationship is clearly an important one 'in a region with extensive uninhabited areas, in which one's horse may have literally meant the difference between life and death" (Dolidze, 1999, p.9), and the connection felt between mountaineer and horse in the Caucasus is probably as ancient as their myths. There is, for example, a Georgian legend that asks, "Who were my ancestors?" And the answer given is "He who pulled milk out of a wild mare's udder with his lips and grew drunk as a little foal". Additionally, "Horses which won special funeral races symbolically took their dead masters to the next world, even in this century in the North Caucasus" (Chenciner, 1997, p.20). Consequently, the fact that the horse plays such a significant role in the story should come as no surprise to anyone familiar with both the geography and history of the region.

The talking horse reminds us that one of the traditional attributes of the shaman is his or her ability to communicate with the animals, and the horse is also frequently the form of transport used by the shaman to access other worlds.

> Pre-eminently the funerary animal and psychopomp, the "horse" is employed by the shaman, in various contexts, as a means of achieving ecstasy, that is, the "coming out of oneself" that makes the mystical journey possible. This mystical journey, to repeat, is not necessarily in the infernal direction. The "horse" enables the shaman to fly through the air, to reach the heavens. The dominant aspect of the mythology of the horse is not infernal but funerary; the horse is a mythical image of death and hence is incorporated into the ideologies and techniques of ecstasy. The horse carries the deceased into the beyond; it produces the "break-through in plane," the passage from this world to other worlds (Eliade, 1964, p.467).

However, the infernal direction, contrary to what Eliade suggests, is not necessarily to the Lower World, just as Heaven is not necessarily only found in the sky. Moreover, the horse in this particular folktale would seem to play a slightly different role to the one Eliade describes. The fact that the youngest son is advised by, and that he consults the horse is an indication of how in *The Sea Horse* it takes on the role of a spirit helper. And the way in which the youngest son is set a number of tasks to accomplish on his journey can be likened to the tasks the shaman sets out to accomplish on the journeys he undertakes.

While on the subject of the importance of the horse, it is worth noting what Eliade has to say on the association between the animal and the shaman's drum, and the way in which the drum has traditionally been used in a number of cultures by shamans to induce the trancelike state required for journeying:

> The iconography of the drums is dominated by the symbolism of the ecstatic journey, that is, by journeys that imply a breakthrough in plane and hence a "Center of the World." The drumming at the beginning of the séance, intended to summon the spirits and "shut them up" in the shaman's drum, constitutes the preliminaries for the ecstatic journey. This is why the drum is called the "shaman's horse" (Yakut, Buryat). The Altaic drum bears a representation of a horse; when the shaman drums, he is believed to go to the sky on his horse. Among the Buryat, too, the drum made with a horse's hide represents that animal (Eliade, 1989, p.173).

According to Yakut beliefs, the horse is of divine origin. In the beginning God is said to have created a horse from which a half horse-half man descended, and from this being humankind was born. The Sky-Horse deity, Uordakh-Djesegei, plays a major role in Yakut religion, and Yakut mythology depicts many other scenes in which deities and guardian spirits descend to the earth as horses. The honourable goddess Ajjyst, the patron of child-bearing, appears as a white mare, as does the goddess called Lajahsit. The horse is of great significance to the shaman too. "A Yakut shaman's healing performance is unthinkable without a horse, just as the entire ceremony cannot occur without the shaman's participation ... A horse, its image, or at times, an object personifying the animal is always present in the shaman's preparations and performances" (Diachenko, 1994, p.266).

As for the reference by Eliade to the horse being known as the Yakut shaman's drum, it is actually the coat the Yakut shaman wears that is believed to give him the supernatural power needed to go to other worlds, and it is this that is called the 'shaman's horse' (see Czaplicka, 2007, p.67).

Among Turkic-speaking people of South Siberia, including Tuvinians, the horse can play an important role too, and it is the drum that can represent the animal ridden by shamans to travel to other worlds. Its handle can be regarded as the horse's "spine; the plaits of leather attached to the upper part of the ring symbolize the reins of the horse; the drumstick is a lash, which beats a drum only in certain places" (Diakonova, 1994, p.253

The style that the storyteller employs is best categorised as that of magic realism, with the unbelievable element being first presented in the form of a dream: "I dreamed of a white sea-horse of rare beauty. When the first rays of the morning sun appeared over the sea, the horse leaped out on to the shore, galloped round the entire world three times and galloped away into the depths of the sea". And through the account of the dream, we learn right from the start what the purpose of the journey will be–for the sons to find this white horse for their father. It is almost as if the father has lost a part of his soul, and the mission for the sons is thus to retrieve and restore it so as to make their father whole again.

Soon after the commencement of the journey, the youngest son is singled out as being special. For it becomes apparent that he has the courage to face up to challenges and to step into the unknown, rather than to look for the easy way out as his brothers do: "Either I will perish or I will come back with success. You stay alive, and tell father of my fate". And this is very much what one would expect of the initiate if he is to be successful when answering the "call".

In the story the youngest son is referred to as a djigit.

Every man and boy in Daghestan carefully keeps and venerates the uniform and arms of his grandfather and great-grandfather, the original *djigits*. The *djigit* was a dare-devil mounted warrior, whose sure-footed horse could float over the terrain, scattered with loose protruding rocks, punctuated with fast deep rivulets and sudden ravines (Chenciner, 1997, p.20).

The shamanic journey frequently involves passing through some kind of gateway. As Eliade explains:

The "clashing of rocks," the "dancing reeds," the gates in the shape of jaws, the "two razor-edged restless mountains," the "two clashing icebergs," the "active door," the "revolving barrier," the door made of the two halves of the eagle's beak, and many more–all these are images used in myths and sagas to suggest the insurmountable difficulties of passage to the Other World [and sometimes the passage back too] (Eliade, 2003, pp.64-65). And to make such a journey requires a change in one's mode of being, entering a transcendent state, which makes it possible to attain the world of spirit (Berman, 2007, p.48).

In *The Sea Horse* this barrier is represented by the dense forest the youngest son is required to pass through, and then the fence of steel stakes surrounding the castle where the giantess lives, on each of which there is a human head. And the fact that there are seven castles in total should also be noted.

Seven is a mystic or sacred number in many different traditions. Among the Babylonians and Egyptians, there were believed to be seven planets, and the alchemists recognized seven planets too. In the Old Testament there are seven days in creation, and for the Hebrews every seventh year was Sabbatical too. There are seven virtues, seven sins, seven ages in the life of man, seven wonders of the world, and the number seven repeatedly occurs in the Apocalypse as well. The Muslims talk of there being seven heavens, with the seventh being formed of divine light that is beyond the power of words to describe, and the Kabbalists also believe there are seven heavens–each arising above the other, with the seventh being the abode of God.

Although the cosmology, described in Creation Myths, will vary from culture to culture, the structure of the whole cosmos is frequently symbolized by the number seven too,

> which is made up of the four directions, the centre, the zenith in heaven, and the nadir in the underworld. The essential axes of this structure are the four cardinal points and a central vertical axis passing through their point of intersection that connects the Upper World, the Middle World and the Lower World. The names by which the central vertical axis that connects the three worlds is referred to include the world pole, the tree of life, the sacred mountain, the central house pole, and Jacob's ladder (Berman, 2007, p.45).

Even at this early point in the plot, the indications are that what we have here is essentially a shamanic story rather than what at first sight might appear to be just a simple fairy tale, and the same can be shown to be the case with many other tales from the region (see Berman, 2007, for two examples from neighbouring Georgia).

The fact that the Narts find the scent of humans objectionable can also be found in other tales from the region (see *Davit* in Berman 2007, for example).

The youngest son then proves himself worthy of being the successor to his father by managing to subdue the white horse: "You have tamed me, and from now on you are my master. Saddle me and tell me where to take you". The horse talks to him with a human voice, and it should be remembered that one of the traditional attributes of the shaman is the ability to communicate with the animals, as the youngest son shows himself able to do.

The fact that the youngest son is advised against picking anything up on the journey lest he fall into the hands of misfortune is also to be expected in an account of what is in effect a shamanic journey. When journeying in other realities, partaking of food is often forbidden, especially when journeying through the Land of the Dead (see, for example, Paul Radin's account of the Winnebago Indian Road to the nether world in the Thirty Eighth Annual Report, Bureau of American Ethnology, Washington, DC., 1923, pp. 143-4, which is reproduced in Berman 2007). In *The Sea Horse*, the prohibition is applied to picking up the feather (getting distracted from the mission), and acts as a reminder of the eristic nature of shamanic practice, often passed over by neo-shamanic practitioners on the workshops they offer. The youngest son, however, cannot resist the temptation and picks up the special feather he finds along the way.

The result is that he is then set the task by the local khan of finding the bird that produced the extraordinary feather. The white horse, acting as his spirit helper, shows him how to. This the youngest son does by refusing to return the plumage to the youngest of three doves who shape-shift into girls when they go bathing. She is thus unable to return to her previous form and the young couple then ride off together.

The young girl's special powers soon become apparent from the way in which she succeeds in tricking the old khan she was supposed to marry, and also from how she manages to rescue the youngest son from the fate his brothers intended for him. She does this by cutting off her tresses and using them to form a rope which he uses to climb up back to the middle world. The story concludes with the young couple getting married. Equilibrium is thus restored to the community once more, which was of course the purpose of the journey.

As an intermediary, the shaman can be said to serve as a bridge or a link–"to facilitate the changing of condition without violent social disruptions or an abrupt cessation of individual and collective life" (Van Gennep, 1977, p.48). And that is what is achieved by the end of the tale.

Furthermore, with the journey involving visits to both the upper and lower worlds (on the back of the white horse), a meeting with a spirit helper in the form of the giantess, being able to communicate with animals, shape-shifting, and retrieving the lost part of a soul, the shamanic elements are plain to see.

> Sceptics will argue that it is impossible to eliminate from analysis the Christian [or Muslim] influence on what sources there are available to us, such that we can never be certain in any one case that we are indeed dealing with beliefs that are authentically pagan. This view is now so widely held that we can in justice think of it as the prevailing orthodoxy (Winterbourne, 2007, p.24).

The same argument could be applied to the attempt to ascertain whether we are dealing with beliefs that are authentically shamanic in *The Sea Horse*. Nevertheless, just because a task is difficult is no reason for not attempting it. If it was, then no progress would ever be made in any research that we might be involved in. For this reason, despite whatever the prevailing orthodoxy might be, there is surely every reason to conduct such a study as this.

In the case of indigenous shamanism,

> the chief methods of recruiting shamans are: (1) hereditary transmission of the shamanic profession and (2) spontaneous vocation ("call" or "election"). ... However selected, a shaman is not recognized as such until after he has received two kinds of teaching: (i) ecstatic (dreams, trances, etc.) and (2)

traditional (shamanic techniques, names and functions of the spirits, mythology and genealogy of the clan, secret language, etc.) (Eliade, 1989, p.13).

In this story the recruitment would seem to be hereditary as it is the youngest son who shows himself worthy to assume the role, and as for the ecstatic teaching, he receives that on the journey, together with an awareness of the powers he has, with the assistance of his "spirits" or helpers.

According to Eliade, "the shaman is indispensable in any ceremony that concerns the experiences of the human soul [when it is seen as] ... a precarious psychic unit, inclined to forsake the body and an easy prey for demons and sorcerers" (Eliade, 1964, p.182).

However, the claim that he / she is "indispensable" would appear to be a poor choice of words as, for example, priests undertaking exorcisms have fulfilled this role too, as have others. On the other hand, the shaman's value in such situations is surely not in doubt and can be seen from the results achieved, and in the case of this particular story this is what the youngest son achieves for his father the *pachakh*.

Eliade then goes on to suggest that the shaman's usefulness in such situations can be attributed to mastery of the techniques of ecstasy:

[H]is soul can safely abandon his body and roam at vast distances, can penetrate the underworld and rise to the sky. Through his own ecstatic experience he knows the roads of the extraterrestrial regions. He can go below and above because he has already been there. The danger of losing his way in these forbidden regions is still great; but sanctified by his initiation and furnished with his guardian spirits, the shaman is the only human being able to challenge the danger and venture into a mystical geography (Eliade, 1964, p.182).

The quote offers further evidence, if any were needed, to support the case that what we have in *The Sea Horse* is the account of a shamanic journey, embedded within what would appear to be a folktale. In other words, what we have here is a shamanic story.

What can also be seen from this tale is that among indigenous peoples, "all concern with health and curing is a religious transaction. If a person suffers from bad health, if he or she falls critically ill, it is all provided for by his or her relations with the supernatural world" (see Hultkrantz, 1992, p.1). Although minor injuries and illnesses such as coughs or colds may not be regarded in this light, generally speaking all disease is believed to have its origin in a disturbed relationship with the supernatural.

Vinogradov (2002) points out how classical shamans among Southern Siberian ethnic groups "always use the spiritual context/vocabulary of a particular culture" and the same observation can be applied to the way indigenous shamans work in other parts of the world too. In other words, as Vinogradov goes on to add, "they are not Universalists like Jungian analysts [or neo-shamanic practitioners], but ... deeply embedded in their [own] cultures and their myths." Moreover, their folktales (and rituals), as we see here, reflect this.

Chapter 3

The Girl-King & The Red Fish

There lived or did not live a khan, and he had three sons. When the khan became old and blind, his sons asked him, "How can you be helped, father?"

"My children! There is only one remedy that will cure me. That is the fruit from the garden of the girl-king. But until now not one man has been able to penetrate into that enchanted garden. It is unlikely that you will manage it either", answered the khan.

"We swear father that we will get those fruits or we will lay down our heads!" exclaimed the sons, and they began to get ready for the journey.

First the eldest son set off. He rode for a long time, until he reached a high snow-covered mountain. Beyond the mountain he saw an old man sewing up the cracked earth. "Salam aleikum! May you not have any success with your stupid enterprise!"

"You yourself will be left without success", answered the old man.

The eldest son struck his horse with his whip and galloped on further. For a long time he was riding and he landed up in a land where a river of milk was flowing, and in the gardens were growing grapes and other remarkable fruits. The khan's son was surprised and thought, "These are probably the very gardens of the girl-king". He filled up his *khurjin* [saddlebag] with fruit and started off on the return journey.

At home he went to his father at once and proudly laid his *khurjin* at his feet. "How quickly you have come back", said the khan in surprise. "Tell me, where did you pick these fruits?" His eldest son related to him about the wonderful land where the river of milk is flowing and the unusual fruits are growing. "In my youth I could gallop to those gardens while the *khinkal* [a Daghestan national dish, similar to dumplings] was boiling", said his father. "No, one cannot get to the gardens of the girl-king quite so easily".

After that the middle son set off on the journey. He also crossed the snowy mountain and met the old man who was sewing up the cracked earth. "May you not have success in your stupid enterprise!!" exclaimed the middle brother instead of a greeting.

"You yourself will be left without success", answered the old man.

The middle son gave no heed to these words and set off further. He reached the land where the river of milk was flowing, and moved on further. He rode thus until he landed up in a land where a river of oil was flowing. In that place there was a cloud of dust, and the mud was up to the knees, but in the dense gardens were growing amazing fruits. "There can be no doubt that these are the gardens of the girl-king", thought the middle brother, and he began to pack his *khurjin* with fruit.

The middle brother also returned home safely and spread out the fruit before his father. "But you have come back quickly, dear son!" remarked his father. "And where exactly did you pick these fruits?" The middle son told him about the distant land where the river of oil was flowing. "In my youth I used to reach that land before my pipe went out", answered the khan. "From the river of oil to the domains of the girl-king is just as far as from heaven to earth".

The turn came of the youngest son. He started off on his way and also met the old man who was sewing up the earth. "Salam aleikum, good old man! Good luck to you!"

"Vaaleikum assalam, *djigit*! Where is your journey taking you?" The khan's son told him that he was searching for the garden of the girl-king.

"Listen to me carefully, I will help you", said the old man. "You will ride through many lands, until you reach the land where a river of honey is flowing. That is the border of the girl-king's domains. She herself lives in a huge fortress with iron gates. You open the gate with a stick that has an iron nail at its end, wrap your feet with grass, go into the garden and knock down the fruit with a wooden stick. Otherwise the iron gates, the thick grass and the trees will kick up a din and you will perish from the sword of the girl-king".

The youngest son thanked the old man and rode on further. He rode for a long time, until he saw a vast palace with stout iron gates. He pushed at these gates using a stick with an iron nail on its end. The gates slowly opened and loudly creaked: "I feel the pressure of iron, I feel the pressure of iron".

"And what other pressure ought you to feel? Do not creak and do not prevent me from sleeping", came the voice of the girl-king. She thought that the leaves of the gates were rubbing one against the other.

The khan's son fastened grass round his feet and entered the garden. "I can feel grass, I can feel grass", the grass began to rustle loudly.

"What can you feel, apart from grass? Do not rustle and let me sleep in peace", said the girl-king, getting angry.

The khan's son began to knock down the fruit with a stick, and the trees began to make a noise: "I can feel wood, I can feel wood".

"Do not make so much noise", the girl-king muttered in her sleep. "Stop knocking your branches together, and you will not feel anything".

When his *khurjin* was full, the youngest son was already about to leave the palace, but suddenly a daring thought came to his head. He carefully made his way through into the bedchamber of the girl-king, kissed the sleeping beauty three times and gently bit her on the cheek.

After that he returned home and gave the fruit to his father. "You have really been wandering for a long time, dear son", said the khan. "When the right time comes, we will try out these fruits".

Meanwhile the girl-king woke up, looked in her magic mirror and saw teeth-marks on her cheek. "Who bit me?" she shouted. The magic mirror told her everything that had happened. "Hurry up and equip the army! I want to have a look at this dare-devil", she ordered. Soon the armies of the seven lands that were subject to the girl-king, took the field.

After some time the huge army with the courageous girl-king at its head made a camp beneath the walls of the town, and messengers came to the khan with a command to deliver up, forthwith, the thief who stole the fruit.

Word of the girl-king and of her extraordinary beauty reached the elder brothers, and they set out first. "It was we who picked the fruit from the magic garden".

"How did you pluck the fruit?" asked the girl-king.

"With these hands", answered the brothers.

"Drive away these liars!" she shouted, and her order was fulfilled at once.

Then the youngest son set out for the camp and he said that he had picked the fruit from the magic garden.

"Tell us, how did you do it?" asked the girl-king. He told her how it all happened. "How did you dare to bite me? I have to pay you in the same coin!" she shouted and she bit the young man very painfully on his cheek. "We are quits", said the girl-king. "But it was not enough. I must punish you yet more. Present the other cheek".

After that the girl exclaimed, "And now take me to your father for his blessing".

The khan ate the fruit and changed into a young man, and when the girl-king touched his eyes he became sighted once more.

In his joy the khan arranged a magnificent wedding. And to his son there soon appeared children: boys like their father, and girls like their mother.

***

Once again, as was the case in the previous tale, the number three has an important part to play-with the khan having three sons, and with the youngest son kissing the girl-king three times. And once again, the two oldest sons show themselves to be unworthy to succeed their father by failing in their quest to cure him of his blindness. The khan's response to the oldest son's efforts is: "In my youth I could gallop to those gardens while the *khinkal* was boiling". And his response to the middle son's failure is similarly dismissive: "In my youth I used to reach that land before my pipe went out". Only the youngest son proves himself to be a worthy candidate, and he does with the aid of the old man he encounters on his journey, the same old man the two older brothers chose to treat with nothing but contempt and disdain. This just shows how misguided they are. For by advising the youngest son on how to journey to the garden of the Girl-King and how to obtain the special fruit there that khan needs, the old man in effect shows himself to be a spirit helper, and it is only because the youngest son recognises the old man for who he is that he is able to succeed.

The Girl-King is clearly a story that features elements taken from a number of different traditions. For example, the "land where the river of milk is flowing" reminds us of "the land of milk and honey" of the Old Testament, as does the youngest son's final destination-the land "where a river of honey is flowing". In other words, what we find here is in effect "the promised land"–a land where everything desired can be realised. Furthermore, the "fruit from the magic garden" remind us of the fruit in the Garden of Eden. The fact that the iron gates, grass, and trees all speak and have feelings in the magic garden is significant too, as this is very much a concept one would expect to find in a culture where animism is practised. On the other hand, the setting is very much an Islamic one–a world where khans are the rulers and people greet each other with the words "Salam aleikum". As for the old man sewing up the earth, he can be seen to represent an attempt to restore the equilibrium of the world that has been upset by the khan's illness, which of course was the traditional role of the shaman.

The *Shimchong: The Blind Man's Daughter* narrative (see Berman, 2007), which is still used in Korea in shamanic ceremonies, illustrates the therapeutic power of storytelling in that the "patient" was supposed to be healed precisely at the climax of the story when Old Man Shim opens his eyes and sees his long lost daughter. We have no way of knowing whether *The Girl King* was ever used for similar purposes but it is a possibility that cannot be discounted.

It has long been known that if a patient is shown his particular ailment is also a general problem–even a god's ailment–he is in the company of men and gods, and this knowledge can produce a healing effect. In ancient Egypt, for example, when a person was bitten by a snake, the priest/physician would recite the myth of Ra and his mother Isis to the patient. The god Ra stepped on a poisonous serpent hidden in the sand that his mother had made and was bitten by it. Knowing that he was threatened with death, the gods caused Isis to work a spell which drew the poison out of him. The idea behind the telling of the tale was the patient would be so impressed by the narrative that it would work as a cure (see Jung, 1977, pp.102-103).

The history of religion can be seen as "a treasure house of archetypal forms from which the doctor can draw helpful parallels and enlightening comparisons for the purpose of calming and clarifying a consciousness that is all at sea" (Jung, 1968, p.33). The message conveyed through the adoption of such

50

an approach is that if the figures we read about and identify with can overcome difficulties, then we can too. And of course contemporary shamanic tales such as *Bundles* (see Berman 2007) can achieve this effect just as well as the ancient biblical stories can.

Tales have long played an important part in the repertoire of healers, who have of course not only told stories but made use of their patients' stories too: According to medical historians, it was only in the 19th century that doctors actually began to examine their patients' bodies in any detail.

> Before that time they would observe them loosely, occasionally take their pulse or inspect their tongue, or peer at their urine or stools. But mostly they relied on the patient's account of what had happened. Eliciting this detailed "history" of their illness, watching for inconsistencies or omissions, and trying to guess at the "true" meaning of what was said, all made medical diagnosis into a type of literary criticism. For a long time, medicine was all about *stories*, not only the patient's "history" and the doctor's "diagnosis", but also the mingling of these two narratives in the medical consultation (Helman, 2006, p.152).

Through the shamanic story, a mythic world can be constructed and symbolically manipulated to elicit and to transact emotional experiences for the patient. As Dow (1986) explains, symbols affect mind, mind affects body and a cure is thus produced by making use of metaphor (see Winkelman, 2000, pp.237-239).

It has been shown through an empirically controlled experiment that "merely telling a human subject about controllability duplicates the effects of actual controllability" (Seligman, 1975, p.48). In other words, when we are told a story, regardless of whether the events in it correspond to the actual state of affairs in the world of experience, it can have the same effect on us as if it were a part of the world of real experience (see Rennie, 1996, p.224). From this it can be seen that the human spirit is not wholly determined by its physical environment but contributes, through the imaginative generation of narrative, to the construction of its own determining environment.

Not only has the shaman traditionally played the role of a healer as the youngest son does in this particular tale, but shamanic practices have also had a considerable influence on contemporary forms of healing. "Specific techniques long used in shamanism, such as change in state of consciousness, stress-reduction, visualisation, positive thinking, and assistance from non-ordinary sources, are some of the approaches now widely employed in contemporary holisitic practice" (Harner, 1990, p. xiii). Jungian and Gestalt therapists also use guided visualisation with their patients to enable them to access inner wisdom. This often involves the patient having a dialogue with an inner sage or teacher in which he / she is encouraged to ask whatever questions seem to be most helpful, and the process can be compared to the shaman's journey to find a spirit teacher (see Walsh, 1990, p.132).

According to Harner (1990), in shamanic terms illnesses, physical or mental, are not considered to be natural to the body and are usually viewed as power intrusions. To resist them you need to be in possession of guardian spirit power and serious illness is usually only possible when a person has lost this energizing force or it becomes depleted. Ingerman (1993), however, maintains there are three possible causes of illness–a person's power animal leaving without a new one taking its place, soul loss, or spirit intrusion. Cases of soul loss are believed to be the result of an emotional or physical trauma. To cope with such an experience, a piece of our life force is said to separate from the body and travel into non-ordinary reality. In psychological terms this is known as *dissociation*. Another cause of soul loss could be the theft of part of our life force by another person.

Being able to put a name to whatever condition a patient is

suffering from can be considered to be part of the healing process too. Naming an inauspicious condition is halfway to removing it. Embodying the invisible in a tangible symbol, such as that of soul theft, can be regarded as a big step towards remedying it and, as Turner (1995) points out, is not so far removed from the practice of the modern psychoanalyst. And once something is grasped by the mind, it can then be dealt with and mastered.

Although emphasis is often placed on the healing of individual illness, either psychological or physical, it should be remembered that another role the shaman can play is in the healing of the community, and the youngest son in our story achieves this by his actions too.

> Travellers to Daghestan have generally commented on the hard life lived by women. During courting, of course, there was a display of feelings of tenderness by the man. Various complimentary descriptions would be made ... After marriage, however, a woman had to work very hard, and generally the older she was the harder she worked, particularly if she became a widow ... Another universal chore for women was the collection of water from the well or spring ... Since the terrain tends to be mountainous and water usually occurs at a low level, this usually meant carrying the full jars uphill. ... In spite of this hard work, it was a woman's ideal to be married ...an Avar proverb: "Light is the grave of a widow without a husband; the sky is the grave slab of a girl without a lover" ...The relations between young men and girls before marriage is strongly governed by the need to preserve the girl's chastity. The normal procedure is for the suitor to send matchmakers to make arrangements with the girl's father (Hunt, 2007, p.2).

In this particular story, however, the girl would seem to have all the power and runs the show, that is, until the youngest son turns up on the scene.

The early representations of women in Daghestan, such as a seventh century B.C. figure that was found in the village of Sogratl, of a naked female charioteer, holding the reins, "show evidence of matriarchal (or at least equal) societies" (Chenciner, 1997, p40) and suggest that Daghestan "was a possible homeland for the legendary Amazons of classical antiquity" (Chenciner, 1997, p.40).   And the tradition of strong women still survives. Daghestan women fought the Russians, both under Shamil and during the Revolution, and today, at work in the villages, they can be seen carrying up to double their own weight on their backs (see Chenciner, 1997, p.41).

***

The Red Fish

This is a fairy tale. There was once upon a time a king who had become blind from old age. The doctors told him that in the White Sea there was a brightly coloured fish with a horn on its head, called the "Red Fish." If it were caught and its blood smeared on the king's eyes, he would recover his sight. The king ordered his son to catch this fish; the prince gathered the fishermen together and they set out.

For two whole days they cast their nets in vain, it was on the third day that they caught the Red Fish. But it was so beautiful that they could not bring themselves to kill it, and so they threw it back into the sea. But the king's son made the fishermen take a solemn vow not to say anything about their catch. Then they returned home.

Now it happened one day that the prince had occasion to beat a negro, one of his father's servants. And the negro ran straight to his master and told him the whole story of the Red Fish. The king was very angry and banished his son from his kingdom. When he said farewell to his mother, she said to him: "If a man follows you

54

on the road, stand still and wait; if he comes right up to you, take him as your companion; if, at dinner, he gives you more then he takes himself, then make friends with him; if he takes upon him to watch at nights while you sleep, pretend at first to go to sleep, then if he really remains awake, be his friend."

Then the king's son said farewell to his mother and went abroad. On the way he saw a man he did not know coming towards him; he did as his mother had told him, and the strange man stayed some distance off. They spent the night in an open field-the king's son pretended to be asleep, but the strange man stayed awake and kept watch the whole night through. In the morning as they were having breakfast the stranger put more before the king's son than he took himself, so that the prince said to himself, "Out of this stranger I will make a good friend."

Soon they came to a town where they lodged with an old woman. "What is the news in your town?" they asked her.

"News is it? Our king has one daughter, she spoke up till her seventh year, but since then she has been dumb. The king has vowed to give her as wife to whosoever can make her speak. But he who tried and does not succeed, his head will be cut off. Many have tried it already-a whole house has been built of their skulls." When the king's son and his friend heard of this they determined to try their luck.

A great crowd assembled in the court of the king's palace to witness the attempt. The friend of the king's son told them to make no answer to three questions he would ask them. Then they all went into the room in which the king's daughter sat behind a curtain. The friend of the king's son began to tell a story.

"Once upon a time a tailor was making a journey. A carpenter joined him on the way, and further on a mullah. They spent the night in a dark wood. The carpenter took the first watch. When he began to get sleepy he took up a piece of wood and carved the figure of a boy out of it. The tailor had the second watch; when he began to get sleepy, he began to make clothes for the wooden figure, in which he dressed it. The mullah took the third watch. When he saw the wooden figure of a boy, and saw that he was fully dressed, he implored God to send the boy a soul. The Almighty heard his prayer, and the wooden figure became a living boy. But in the morning the three began to quarrel; each of them claimed the boy. 'He belongs to me,' said the carpenter. 'No to me,' cried the tailor. 'What are you thinking of? The boy is mine!' said the mullah. Now, good people, what do you think? You who are gathered together here, tell me, to which of these three should the boy belong?"

But no one answered; even when the teller repeated his question, everyone still remained silent. Only the king's daughter could not stand it any longer. "Why do you not answer?" she cried behind her curtain. "The boy belongs, of course, to the mullah!" Immediately the whole people sprang joyfully to their feet-"Good! She has spoken," they all shouted together. And the king gave the prince his daughter to wife.

At night when the bridegroom was about to go to his bride, his companion told him not to lock his door. And when the young couple were sleeping, the friend went in and saw that an enormous serpent was crawling into the room. He killed it with his diamond sword, and when day came everyone saw what had happened through the night.

Ten days later the son-in-law left the palace for his own home. The king gave him ten servants and gave his daughter ten slave-women as well as ten camels with ten loads of costly goods.

When they came to the place where the stranger had attached himself to the king's son, the stranger said to him: "Now we must divide everything between us." The king's son was well pleased with this arrangement and they divided everything, the goods, the servants, the slaves. Only the king's daughter remained. "We must split her in two," said the friend. "No, no, do not kill her! Rather take her altogether," said the king's son. But in vain: the other refused to hear of it. So they bound the king's daughter to a tree, the friend drew his diamond sword and pretended he was going to split her head. But she was so terrified that she became sick and ... little snakes came crawling out of her mouth. The stranger swung his sword a second and a third time and then unbound the king's daughter.

"A serpent fell in love with her," he then said to the king's son, "and slept beside her every night. Then king's daughter became dumb from breathing the serpent's breath and was about to give birth to these little serpents. Now I must leave you. I present you with my share. Your father is blind; take a little earth from the hoof of my horse, smear it on his eyes, and the light will return to them. You will not see me again. I am the fish you would not allow to be killed." He had hardly spoken when he had already vanished.

But the king's son went home with all that he had, with servants, slaves, camels, costly goods and his young wife. He smeared his father's eyes with earth from the hoof of his friend's horse, and at once they received back their sight.

And now ... our fairy tale is done.

***

The Red Fish is a Zachurian story, taken from Dirr, A. (1925) *Caucasian Folk-tales*, London: J.M. Dent & Sons Ltd. (translated into English by Lucy Menzies). The Zachurians are described in the Introduction to the volume as being a Lesghian race living in Upper Samur.

The poor health service available in Daghestan no doubt helps to maintain the still widespread practice of folk medicine:

Spring waters, herbs and spells form an integral part of this medicine ... The curing spring beside the waterfall at Inkhokvari, near Aguali village is one such mysterious example. When the moon waxes, there is more gas in the pungent waters, when moon wanes, less, and the flavour changes. Here I met ailing Jakhfar Gazi Magomedov, whose sons had brought him 200 kilometres to cure his stomach by drinking the waters. They then prayed on mats by the river. Earlier this century, on the night of the equinox, even ordinary spring water was considered to have medicinal properties by the Kumyks, who either bathed in the river or brought home ewers filled with its water to pour over themselves and their elderly relatives. They used special brass bowls, with central raised bosses and magical inscriptions, for drinking remedial waters (Chenciner, 1997, p.85).

Whether the storyteller, collector, or translator chose to place "This is a fairy tale" at the start is not clear, but at first sight it would seem to be rather unfortunate as this colours the way we react to it as listeners or readers.

The Red Fish, like *The Girl-King*, is a healing tale in which the mission for the journeyer, the king's son, is to find a particular red fish with a horn on its head, the blood of which can cure blindness if smeared over the sufferer's eyes.

Although the son locates the fish, he does not have the heart to kill it so informs his servants to keep the fact a secret. The king finds out however, and as a result he banishes his son.

On his journey into exile, the king's son is befriended by a stranger. They lodge together with an old woman. There they learn about the local king's daughter, who has been struck dumb, and will be given as a wife to anyone who can cure her.

The stranger advises his companion on what he needs to do to get the local king's daughter to speak again, and the plan works. It involves telling a riddle, the question at the end of which the princess feels compelled to answer. The riddle itself is of interest as it is about a figure carved by a carpenter, a mullah and a tailor while they are feeling sleepy, in other words in a trancelike state, that is given a soul, and the question is who the soul should belong to.

Not only does the stranger bring the young couple together, but he also rescues them on their wedding night from an enormous serpent.

On their way home, the two friends agree to share equally everything they have been given by the grateful king, though they have a problem with how to divide the daughter. The suggestion is for the stranger to cut her in two, which scares her so much that she vomits, and little snakes come out of her mouth. The cause, the stranger informs his friend, is that a serpent fell in love with her, something that can perhaps best be described as a case of soul captivation.

The stranger, who turns out to have been the very same fish that the king's son once saved, then vanishes, and the son returns home again, where he restores his father's sight by smearing his eyes with earth from the hoof of his friend's horse. So the son, taking on the role of shaman, heals his father, with the shape-shifting fish acting as his spirit helper.

Then we are informed "our fairy tale is done". However, the effect of telling us once again that what we have heard is a fairy tale is to make us think that perhaps what we have heard is in fact something else, and possibly this was the storyteller's intention all along, which would also explain why it was introduced the way that it was.

Chapter 4

The Black Fox and Sartanki

There lived or did not live a certain poor man. He had a son who was reputed to be very fair and just. The father and son took it in turns to pasture the *aul's* cows and for that they received a meagre remuneration. (An *aul* is a type of fortified village found throughout the Caucasus Mountains, especially in Daghestan). One day the son drove the herd to the seashore and saw a huge fish gasping for breath on the sand.

The herdsman beckoned to a village acquaintance and asked him to send fifteen carts in order to load them with the meat of this fish. As soon as the messenger went off, the fish began speaking with a human voice and asking the herdsman to preserve its life. "Help me to reach the water, and in a difficult hour I will help you.

"How will you be able to help me?" asked the herdsman.

"Here, cut off my whisker and, when you are in need of help, burn it right on the shore. Then I will swim to the shore", said the fish. The young man agreed, cut off the whisker and helped him to reach the water.

A little later his father came with a cart and the neighbours. "Where is your fish?" asked the old herdsman. And his son told him all that had happened. "Ach, you wretched miserable liar! Where was it ever seen that a fish spoke with a human voice?!" His father got angry and began to wallop his son.

The young man just managed to tear himself away and to take to his heels. For a long time he was running, swallowing his tears, until he reached a dense forest. "Let happen what will!" exclaimed the young man, and he went into the thickest part of the forest.

Here he saw a large deer that had fallen onto a strong trap. The herdsman was just about to kill the deer, but the latter started speaking in a human voice: "Do not kill me! In return I will help you when you really need me. Pull out one of my hairs, and when you burn it I will come running at once".

The young man freed the deer, plucked one of his hairs and went on further. A little later he heard a pitiful cry. It was a big eagle crying, and a snake was creeping towards its nest. Without delay, the young man killed the snake.

"Thank you, kind man", the eagle thanked him. "At a difficult moment I will help you. Here, take my feather. Burn it, and I will come flying". The young man took the feather and went on further.

Soon he left the forest and found himself on the steppe. Suddenly the young man saw a black fox who was being pursued by a pack of dogs. They were overtaking the weakened fox and ready to tear him to pieces. The fox ran up to the young man and said in a human voice, "Save me, traveller! In return, I will help you at a difficult moment. Pluck a hair from me. You only have to burn it and I will come running".

The young man drove off the dogs, plucked a hair from the fox and moved on further. Towards evening he reached some town, knocked on the first house he came to and asked for lodging for the night. The host let him in, fed him and made a bed for him.

In the morning the young man went to have a look at the town and he saw a big palace. Around the palace was a palisade, on which human heads were stuck up.

The young man asked his host about it, and he related, "These are the heads of suitors who have been seeking the hand of the khan's daughter. She is a famous beauty in our region. But she can only be married by the one who is able to hide himself from her penetrating gaze. And no matter where the poor suitors hid themselves, each time the girl found them and killed them without pity".

"I have to obtain the hand of this heartless girl!" exclaimed the young man.

"Do not go to certain death", his host was trying to talk him out of it. "I am sorry for you; after all you are still a child".

"So much the better, small people are more difficult to spot", laughed the young man, and he went to propose to the khan's daughter.

When he arrived at the palisade, he started shouting loudly, "Eh, khan, get ready for the wedding, the bridegroom has come!"

The khan and his daughter climbed up on to the palace roof, looked at the young man and said, "All right, bridegroom, hide yourself; and well enough so that nobody can find you".

The young man went to the sea, pulled out the fish whisker and burned it. Right away the huge fish came swimming to the shore. "Give your orders", said he. "I have long been waiting for you".

"Hide me, and well enough so that nobody can find me", he ordered.

The fish opened his mouth and swallowed down the young man. Then he swam to the bottom and buried himself in the sand.

But the girl had a look in all four directions: there was no sign of the young man on the earth, nor in the sky. She looked at the sea and saw the young man in the belly of the big fish. "Come out of his belly and hide yourself a bit better!" shouted the khan's daughter.

The young man climbed out of the fish's belly, took out the deer's hair and burned it. Immediately the deer presented himself before him. "Give your order, I have long awaited you".

"Hide me so that not a single man could find me", ordered the young man.

"Jump on me", said the deer and he hurtled him away into the very densest thicket of a forest which was beyond seven mountains.

But the girl went out on to the roof, looked in all four directions and saw him in the forest thicket. "Come out of the forest and hide yourself still better!" she shouted.

The young man pulled out the eagle feather and burned it. At once the eagle came flying and said, "Give your order. I have long been awaiting this hour".

"Hide me, and so well that the khan's daughter cannot find me", ordered the young man. The eagle sat him on his back, shot upwards, flew up to a black cloud and began to soar there.

The khan's daughter went out on to the roof, looked at the sky and began shouting, "Climb off the eagle and come into the palace to me!" The young man, hanging his head, dragged himself to the palace.

"Are you ready for death?" the khan asked him.

But the young man remembered the black fox and asked to be tried a fourth time.

"Where will you get to the fourth time?" laughed the khan. "Nobody has ever yet evaded my daughter's gaze".

"Let the young man try for the last time", asked his daughter.

The khan was willing, and the young man left the palace. Beyond the first corner he pulled out the fox hair and burned it. The fox appeared instantly and, wagging his tail, he asked, "What is your order, *djigit*?"

The young man told him everything and asked. "Hide me from the eye of the khan's daughter. Otherwise I will have to perish".

The fox started whispering something, and he turned himself into a morocco-leather trader, and he turned the young man into a flea. "Now I am going into the palace and will sell my morocco leather", said the fox. "And you jump over on to the girl and hide yourself under her arm". And soon the morocco-leather salesman knocked on the palace gate.

"Who is there?" responded the gate-keeper.

"It is I, the morocco-leather merchant, and I am carrying

goods for the khan's daughter".

"Let him in", ordered the khan's daughter, when she was told about the merchant.

While she was choosing some goods, the flea inconspicuously jumped on to her and hid himself beneath her arm. Then the girl let the trader go, climbed up on to the roof and began to spy out the young man. She looked in all four directions: he was not there. She looked in the sky, and he was not there. She looked in the sea, and she did not see him there either.

"Look thoroughly, dear daughter", requested the khan.

"No, father. I cannot find that cunning fellow", admitted the girl finally.

And as soon as she had pronounced these words the flea jumped out and turned himself into the young man. The khan was forced to give away his daughter to the herdsman's son and to arrange a feast that was fitting to the honour of his house. For three days they played the *zurna* [a double-reed wind instrument], beat on drums, and fired guns. I left the gaiety and quietly walked out, in order to tell you all about it.

***

The Black Fox can best be described as being a story about communicating with animals, and shape-shifting.

Shape-shifting can be viewed as the imitation of the actions and voices of animals, though shamans themselves would certainly not describe what they do in such terms. During his / her apprenticeship, the future shaman has to learn the secret language that is required to communicate with the animal spirits and how to take possession of them, and this is often the "animal language" itself or a form of language derived from animal cries. It is regarded as equivalent to knowing the secrets of nature and hence evidence of the ability to be able to prophesy.

By sharing in the animal mode of being, the shaman can be

seen to be re-establishing the situation that existed in mythical times, when man and animal were one (see Eliade, 1989, pp.96-98).

There are many accounts of the incredible feats supposedly performed by shamans. For example, it was said of the Lapp shamans, the *noiaidi*, that as well as having the power to summon herds of wild reindeer, they were also able to transform themselves into animal forms–such as a wolf, a bear, a reindeer, or a fish (see Ripinsky-Naxon, 1993, p.63).

According to Philip "Greywolf" Shallcrass, described as a Druid shaman, shape-shifters, also known as "theriomorphs", can be found in both the Irish and the Welsh traditions too. *Fintan mac Bochra*, for example, regarded as a fount of wisdom by the Irish *filidh*, transforms into an eagle, a salmon and a stag. And in the Welsh *Mabinogion*, *Lleu Llaw Gyffes* transforms into an eagle, while other characters become deer or wolves. Then there is the Story of Taliesin, which has the bard and the goddess Ceridwen both going through a series of animal transformations (cited in Wallis, 2003, pp.86-87).

The belief in a human's ability to change into an animal form also flourished in Renaissance Italy. Among the authors of the many learned and respected books that were published on the subject was one by a friend and colleague of Galileo, Giovanni Batista Porta. His book *Naturall Magick* described the process of metamorphosis by the use of psychotropic substances (see Ripinsky-Naxon, 1993, pp.63-64).

The Brothers Grimm fairy tale *The Three Languages* is all about learning the language of the animals, and then showing how it comes in useful–the way the two white doves teach the new Pope to say Mass, for example. And if we take a view of reality in which everything is understood to be inhabited by a spirit similar to all other spirits, there is no problem in believing that man can change into an animal, or the other way around (see Bettelheim, 1991, pp.46-47).

Through shape-shifting shamans can be said to be "identifying themselves with the very powers that deeply threaten them, and ... enhancing their own powers by the very power that threatens to enfeeble them" (Turner, 1995, p.174) and this parallels what children unconsciously do when they play games.

The earliest example of a humans taking on an animal role in Daghestan

is from a 14th-century worn carved stone relief from Koubachi village, which appears to portray a man, rushing on clawed bird-like feet, wearing a *simurgh*-bird head mask with long feathers streaming behind. The wavy serpent coiled in submission around his waist reproduces the other partner in this Zoroastrian cosmic conflict. This was the religion of the Sassanians, whose Iranian emperor, Khosrows, built the walls of Derbent, just a day's ride from Koubachi, down to the Caspian Sea (Chenciner, 1997, p.196).

Shape-shifting can still be seen to be taking place, even in our own times, in the way shamans have evolved and survived into the 21st[t] century by developing preservation strategies. This has enabled them to adapt to the new situations they find themselves in and so remain relevant to the communities they serve.

As has already been mentioned, one of the attributes traditionally associated with the shaman is his or her ability to communicate with the animals and for them to act as helpers, and in this particular tale the fish, the deer, the eagle, and the black fox all speak to the herdsman with a human voice. Moreover, in return for his kindness, all offer to assist him by providing a whisker, a hair, or feather for him to burn should the need arise.

When the herdsman takes up their offer of help, it is the fox who ultimately succeeds in outwitting the khan's daughter, and this is what has been written about what the animal represents in Daghestani folklore:

The fox is agile, bright and brave, so eases out of awkward situations and wins out over the stronger animals, such as wolves, bears, leopards and lions. The fox stands for people who are swift to recognise social injustice. So, in a struggle with her stronger enemies the fox is sure to win, whereas weaker opponents always triumph over her. The fox is crafty, not only fooling animals but people as well.

In a tale called 'Winter is ahead, I will still catch up with it', the fox is on her way to see the wolf and meets some girls who surround her and ask her to dance the *lezginka*. The fox agrees but insists she must be dressed for the part and they adorn her with all their kerchiefs, rings, bracelets, earrings and belts. The fox dances and asking the girls to make a bigger circle, slips out and runs away to the stream where the wolf has been fishing with his tail. On seeing the fox so smartly dressed, he gladly exchanges his catch for the stolen finery.

Naturally, folklore passes into language and sayings originate from animal stories, for example "he has a fox's walk," and "a serpent whispers but a fox talks." Daghestanis consider that the craftiness of a fox lies not in her head but in her tail, and so boast "if you're a fox, then I'm the fox's tail" (Chenciner, 1997, p199).

We learn that the khan's daughter "can only be married by the one who is able to hide himself from her penetrating gaze". The evil eye is an ancient and widespread belief that certain people have the power to harm or even kill just with glance, and that various means, such as the use of charms or gestures, could be employed to counteract its effects. For example, Virgil speaks of an evil eye bewitching lambs: "Nescio quis teneros oculus mihi fascinat agnos" (*Bucolics, Ecl.* iii, 103).

The evil eye was always held to be capable of bewitching and had to be avoided, as it could cause illness or even death. Even today, beads to keep the evil eye at bay are sewn on children's clothing from birth, or worn around their wrists or necks. Small glass beads, resembling eyes, were favoured by the Dargins, as were the bones or teeth of boar, fox, wolf or goat, sea shells, bears' claws, and egg shells. The Lezgins protected themselves with amulets in red cloth bags or leather pouches, adapted in Muslim times to hold a few lines from the Koran, sewn into their clothing. Richer folk would keep an amulet box, containing barley, apricot kernels, quince or cornelian cherry, as fruit-bearing plants were thought to have healing properties (Chenciner, 1997, p.86). .

The herdsman, during the course of the tale and once he overcomes the barrier represented by the palisade around the palace on which human heads are stuck, journeys from his starting point in this reality, first to the bottom of the sea in the belly of a fish, then to a forest beyond seven mountains, up to a black cloud in the sky, and then he shape-shifts into a flea before finally resuming his human form once again. And if this is not a description of a shamanic journey, then it is difficult to know what would be.

***

Sartanki

The Lak story that follows was taken from Dirr, A. (1925) Caucasian Folk-tales, London: J.M. Dent & Sons Ltd. (translated into English by Lucy Menzies). The Laks or Ghazimukcks are described in the Introduction to the volume as being "of Lesghian extraction". *Sartanki*, like *The Black Fox*, is about shape-shifting, which is why the two stories have been placed together.

***

There was once, there was once, there was once a king, and he had a son. Now this prince went with his servants one day to the hunt. And when a stag crossed over in front of them, the prince made after it as fast as his horse could carry him. But one by one the servants fell behind. Suddenly the stag disappeared into a cave. As it had grown dark by this time the prince lay down to sleep. While he slept the stag stole softly out of the cave, and laid sugar and straw in the rim of the prince's hat. In the morning the prince got up, mounted his horse and set out home. He met his servants one after the other; they all rode homewards together, but when they got there the prince was suddenly taken ill and had to go to bed.

He was so very ill that after a little time it seemed he was going to die. It was in vain that his father brought doctors from far and near, no one could cure him. When the prince felt his end was near, he begged his father to let him be carried to the bazaar. So he was laid on his bed, covered with a coverlet of silk from Schemache, and carried to the shore of a little lake beside the bazaar [In Kumuch, where the story was told to the collector, there was apparently a large pond in the vicinity of the bazaar]. As he lay there, an old bald-headed man passed by, looked at the prince and said: "Look! That is he who has fallen so deeply in love with Sartanki!"

The prince's attendants asked the old man if he could not cure the prince. "Certainly, I can," he answered, "what should prevent me?" The servants at once ran to tell the king the joyful news. He had the old man brought to him and asked him how he intended to begin to cure the prince. "First of all, look at the straw and the sugar in the brim of the prince's hat," said the old man. The servants went to look and were not a little surprised when the articles mentioned were found. The king then asked the old man what was to be done next. "If you marry him to Sartanki, he will be cured," was the answer. "If not, then he will die. Give me those things which I will ask of you and then let the prince come with me: I will take him to her whom he loves." The king ordered that the things he asked for should be given to him; Bald-pate took them and disappeared. After a week had passed he came again, with two horses: he mounted one of them himself, the prince leapt on the other, and they rode away together. And so they rode till they came to the shore of a certain sea. "What are we to do now?" asked the prince. "How are we to get across?" "Do not disturb yourself," answered Bald-pate, who pulled out a net and gave it to the prince. "Put that over your eyes, prince," he said, "we are going to ride through seven seas. On the bottom of these seas you will notice many beautiful things, pearls, diamonds, corals, gold and silver. But touch nothing! Let everything remain as it is."

So they crossed the first sea, and the second, and the third, the fourth, fifth, sixth and seventh, which was the last. Then Bald-pate took the net from the prince's eyes and put it back in his pocket. They went on further and further till they came to a town, where they lodged with an old woman.

"Will you take us God-sent guests into your house?" asked Bald-pate. "If you are sent from God," she replied, "then may I serve as a sacrifice to you. Why should I not receive you? But I have nothing to eat or drink. All that I can give you is an empty room. If that will serve you, then come in."

Bald-pate put his hand in his pocket and pulled out a handful of gold which he gave to the woman. Ah! How overjoyed she was! How she sprang about the room! And then she led her guests to another room, which was handsomely furnished, and served them with such a great dish of pilaw that they could not see over the top of it. After they had eaten, Bald-pate called to her: "Well, old woman, what is the news of your town? Is there justice and order here?"

"Both of these," she replied, "but there is one thing that is not good. Our king has one daughter, Sartanki; she can change herself into any kind of creature that suits her, but she will not marry."

"Can you take us to her?" asked Bald-pate.

"Why not?" said the old woman. "I go to her every day, to comb and dress her hair."

The Bald-pate pulled another handful of gold out of his pocket and gave it to the old woman, who knew at once after that how to arrange matters. "Tomorrow morning," she said to the prince, "I must go to the princess. Take a golden samovar on your shoulder and follow me. And when you arrive in front of the castle, call out your samovar as if you were a dealer.

And so it all happened. The old woman went to the castle and combed and dressed the princess's hair, the king's son came with the golden samovar and offered it for sale before the castle. When the old woman heard his voice, she looked out of the window and called the princess to her. "Look, Sartanki, look at that youth down there, see how handsome he is! If someone else took him for a husband!" Then the princess gave orders that the young man should be called up, and he had hardly come into the room before she recognised him as the huntsman who had driven her into a cave when she had changed herself into a stag. The old woman at once disappeared from the room.

"Listen," said the king's son to the princess, who was lying in his arms. "Your father will certainly not give you willingly to me as my wife. The best thing we can do is to run away." The maiden agreed to this, and after a time begged her father's permission to go away for three days' hunting. The father had no objection to that; she met her lover and Bald-pate in the wood in which she had announced she was going to hunt, and all three rode off on the prince's trusty steeds towards his home.

But when the three days were up and the princess did not return home, her father began to get suspicious. He looked first in her room; it was locked, and when it was broken open it was found to be empty.

"That old woman is probably at the bottom of this!" said the king; he summoned her to him and asked her if she knew where his daughter was. The old woman pretended to know nothing, and only when the king got out his riding-whip and struck her black and blue did she confess the truth. The king was in a great rage and determined he would not leave one stone of the kidnapper's town standing on another. He assembled his army and set out to bring back his daughter and to slay the king's son and all his house.

In the meantime the king's son with his bride and Bald-pate had drawn near to his native town. On the way they met an old man who was walking up and down, laughing and crying by turns. "What is the meaning of this?" asked the king's son; "why are you walking up and down, laughing and crying?"

"Sire," he replied, "the son of our king has died in foreign lands and this is the day set apart to honour his memory. I weep because our prince is dead, and then I laugh when I think of the gifts that are to be given away."

"Well, well," said the king's son. "I am he who is reported to be dead. Now run to the castle and tell my father that I am alive and coming to him. Run quickly and he will give you a good reward."

The old man did not wait to be told twice, but ran as hard as he could with the good news. The king went out with all his viziers and ministers to meet his son and arranged a splendid wedding for him and Sartanki. But when, in a little time, the bride's father approached with his army, the king sent him a message that there was no need to declare war, he had married his son to Sartanki according to the commands of Islam, and if he cared he might attend as a guest. And so it happened; he stayed for three days in his daughter's house, and then in a friendly and happy spirit said his farewells and went home to his own country.

***

The style of storytelling employed is once again that of magic realism, a fusion of logic and "nonsense", in which the logic is satisfied by the factual information provided: "So he was laid on his bed, covered with a coverlet of silk from Schemache, and carried to the shore of a little lake beside the bazaar".

Alternatives are what we try when all normal courses of action prove to be ineffective: "It was in vain that his father brought doctors from far and near, no one could cure him". It is at this point that the old bald-headed man appears to take on the role of the prince's spirit helper, to explain the cause of his condition, and also how a cure can be effected.

As has already been mentioned, the shamanic journey frequently involves passing through some kind of gateway, and in this particular story the seven seas represent the barrier between the two worlds. Also, when journeying in other realities, the partaking of food is often forbidden, especially when journeying through the Land of the Dead (see, for example, Paul Radin's account of the Winnebago Indian Road to the nether world in the Thirty Eighth Annual Report, Bureau of American Ethnology, Washington, DC., 1923, pp. 143-4, which is reproduced in Berman 2007). In *The Sea Horse*, the prohibition was applied to picking up the feather, and in *Sartanki* it is applied to touching anything along the way. As for the number seven, its significance is referred to in Chapter 2, in the commentary on *The Sea Horse* (see p.40).

The mission of the journeyer, on this occasion is a personal one rather than one on behalf of the community, to locate and return with the object of his affections. The old man encoun-tered on the return journey "walking up and down, laughing and crying by turns", appears to be a kind of trickster figure, and is indicative of the vacillation between the two extremes of emotion that tends to take place in liminal states before reincor-poration is achieved. In *Sartanki* the equilibrium is restored on the bride's father acceptance of his daughter's marriage.

How could the old bald-headed man have known that the prince had fallen in love with Sartanki, and that this was the cause of his malaise? The answer must surely be that he was gifted with the power of divination, which has been defined as "a means of discovering information which cannot be obtained by ordinary means or in an ordinary state of mind" (Vitebsky, 2001, p.104). This would suggest that an astrologer is not in an ordinary state of mind when he constructs a chart and a palmist is not in an ordinary state of mind when he reads the lines on someone's hand. However, there is no reason why this should be the case. Consequently, Harner's definition is to be preferred: "Shamans are especially healers, but they also engage in divination, seeing into the present, past, and future for other members of the community" (Harner, 1990, p.43). To the word "seeing", however, I would add "with the help of the spirits" for as indigenous and neo-shamans alike have pointed out, they do not make use of their own power for this purpose.

The shaman as Diviner may use the method of possession, enter a trance-like state, interpret omens or signs found in events in nature, cast stones, sticks or bones, or he may even look for the patterns in the markings on the liver or the shape of the entrails of an animal. Whatever means he chooses to employ, the under-lying belief is always the same—"that the whole universe is inter-connected and has a common pattern running through it, so that if the skilled person looks carefully at any one part of it he will be able to read off what is happening in other parts" (Turner, 1971, p.35).

We know that pagan priests in Daghestan "used to prophesy by examining the offal and other internal organs of the sacrifice" (Chenciner, 1997, p.119), and we also know that such methods of divination were prevalent in other cultures where forms of shamanism were practised.

Chapter 5

The Khan and his Two Sons

Once there lived an intelligent and just khan. He had everything: extensive pastures, cattle, and beautiful fields. The only thing he did not have was children. As he was getting older, the khan would more and more often become sad that he had no son who would inherit all these riches.

One day on the road he met an unknown man, who said to the khan, "I know what you are sad about. I will help you in your grief, if you will fulfil my condition.

"Help me, and I will agree to everything", answered the khan joyfully.

"Take these two apples, give them to your wives and they will bear you two sons. However, when they have reached the age of fifteen, you must send one of them to my castle", said the unknown man, and he vanished.

The khan did just that. And so his wives bore him two handsome golden-haired sons, as like one another as two drops of water. When the sons reached the age of fifteen, the khan called them to him and told them about his past agreement.

"And so the time has come when I have to send one of you to him, and that man is a wicked Nart who kills all of those he meets", the khan concluded his story.

Each one of the brothers began to insist that he is the one to be sent. Finally the khan rested his choice on the son born from the ordinary woman.

Before his departure the doomed son shot from his bow into the ceiling and said to his brother, "If from the arrow there drips milk, then I am alive, but if it drips blood, then you must know that a misfortune has happened to me".

The young man set off to the wicked Nart. Along the road, in a dense forest he met a venerable old man. "Salam aleikum, kind man!" the young man greeted him.

"Vaaleikum assalam, *djigit*. Where is your journey taking you?"

The khan's son told the old man everything, and he answered. "I know that Nart. He is an evil magician. But there is a remedy even against his magic. When you arrive at the Nart's palace, he will ask you to chop wood, to gather firewood and to light the oven. Do not fulfil any one of these requests. And when he himself climbs into the oven, push him into it with a stone. Death will then overcome even the Nart".

The khan's son thanked the old man and soon came riding to the Nart's palace. The magician received his guest well, fed him and put him to bed. In the morning the Nart went into the forest and came back with a huge tree. "Come on, would you mind cutting up this tree!" he shouted.

"Whoever heard of guests working?" said the khan's son, refusing.

The Nart angrily struck the tree against the floor, and it was smashed into splinters.

"How about gathering up the firewood!" ordered the Nart.

"I won't even consider it. Guests must rest for three days".

"Come on then, light the oven!" the Nart gave him a third order.

"Light it yourself!" answered the young man.

The Nart climbed into the oven, and the khan's son blocked up the opening with a stone. A light little cloud rose up: that was the Nart's soul leaving his body.

After that the young man began to open the doors of the palace. In five of the rooms he found he found human skeletons, in the sixth a wonderful horse, and in the last one a beautiful girl.

The overjoyed girl asked her liberator to stay with her in the castle, but the *djigit* sat on the wonderful horse and rode on further. Meanwhile he had told the girl to remain as mistress of the castle until his return.

Whether his journey was long, we do not know, but after a little time the *djigit* reached a large town. Here he met the khan's herdsman. "Accept me as a herdsboy", the *djigit* asked him. "I am a good worker, and what is more, I do not need a big remuneration". The herdsman agreed.

And so a little time passed. In the autumn, when the flocks were driven into town for counting, the shepherd and his herdsboy sat down beneath an apple tree in the palace garden. By coincidence the khan's three daughters called into the garden. They were carrying on their shoulders silver pitchers of water. "Let us quench our thirst, girl", the herdsboy asked the khan's youngest daughter.

"What an impudent fellow!" exclaimed both of the elder daughters, and they called the shepherd names.

But the youngest sister, in defiance of her elders, let the herdsboy quench his thirst.

"Thank you, my beauty", he thanked her and raised his cap. The khan's youngest daughter saw his golden hair and fell in love with the handsome youth.

"Let us give father a present, we will each pluck an apple", suggested the eldest sister. The girls each picked an apple and took it to their father.

The khan was very pleased by the attentions of his daughters. The apple that the eldest daughter had brought turned out to be entirely rotten, the apple of his middle daughter was beginning to rot, and only the youngest daughter's apple was ripe and tasty. "How amazing!" exclaimed the khan, and he asked his wise men to explain this surprising coincidence.

All of the wise men exclaimed with one voice, "O powerful khan! Your eldest daughter has obviously become past her prime, the middle one is close to it, and only your youngest is in the flowering of her youth. Do not waste time, our khan, in finding husbands for your daughters".

"It is wise advice", agreed the khan and he told his daughters to select bridegrooms. The eldest pointed to the son of the vizier, the middle one to the son of a *nazir* [a local figure of authority], but the youngest one said that she would marry only the khan's herdsboy. The khan was distressed. He started trying to dissuade her, but the girl was adamant and the khan was obliged to give his daughters to their sweethearts.

Soon the youngest daughter stopped going to the palace. The retainers tormented her with their gibes, and even more so did her elder sisters and their husbands.

After a little time the khan became seriously ill, and the doctors declared that only the meat of a white djeyran [a gazelle, a springbok] can save him from death.

The sons of the vizier and of the *nazir* set out at once to look for a djeyran. Meanwhile the herdsboy changed his clothes, sat on his wonderful horse and also galloped off to hunt. He easily caught up with a white djeyran, killed it, tied it to his saddle and started riding home.

Along the road he met the sons of the vizier and the *nazir*, who had not been able even to see a djeyran. "Sell us a little of the meat, *djigit*", they asked the rider, not recognising the khan's herdsboy.

"I will give you some meat, but in return I must mark you with my *kleymo*" [a brand. Branding is a motif often found in the oral literature of the people of Daghestan. In the Middle Ages in Daghestan there were both *lagov* (slaves) and *rayatov* (semi-dependent persons), and signets were used to mark their ownership].

The proud fellows did not want to carry a slave's *kleymo*, but they were afraid to go back with empty hands and to suffer the mockery of the khan's retainers, so they agreed.

The herdsboy put his *kleymo* on them and gave them the worst pieces of meat, which he surreptitiously smeared with bile.

The sons of the vizier and the *nazir* returned home and told their wives to take the meat to their father. As soon as the khan tasted the meat he spat it out indignantly and shouted, "That is not the meat of a white djeyran, but that of the most scraggy goat! Evidently good hunters have become extinct in my khanate".

Soon the youngest daughter entered the palace and offered the meat of the white djeyran to taste.

"Where did you get it?" said her father in amazement.

"It was my husband who brought it", answered his daughter.

"How could your pauper manage to get it? He also has probably slaughtered an old goat?" said the khan, but nevertheless he ate the meat and felt better immediately. Soon he had completely recovered.

Some time passed, and a new problem arose. A countless horde of enemies fell on their khanate. The khan collected a large army and set his vizier's son at its head, and appointed the *nazir's* son as his assistant. The herdsboy also wanted to go to war, but the leaders drove him away with ridicule.

The battle had gone on for three days, and the enemy army was beginning to win. Only the night saved the khan's weakened army from complete defeat.

Meanwhile the herdsboy's wife spent all night sewing her husband clothes of red cloth, and in the morning he galloped off to the field of battle. At the very climax of the fighting, the enemy ranks faltered: that rider in red was sowing death and terror among them. Finally the enemies retreated.

The triumphant conquerors returned home, and the khan in his joy arranged a big feast. "But who exactly was the golden-haired horseman in red? Search him out", ordered the khan. But the *djigit's* trail had gone cold.

On one occasion the wives of the vizier's son and the *nazir's* son decided to visit their youngest sister. They did not find her at home, and out of boredom they began to rummage in the corners. Suddenly, beneath a heap of clothes, they saw the red *cherkeska* [a Caucasian coat with a tightly fitted frogged upper part and skirtlike lower part] "Ach, so this is who he is, the man in red!" they exclaimed, and they ran at once to tell their husbands about it.

"For us it will be a great disgrace if everybody knows that we could not overcome the enemy army without the help of this pauper", said the vizier's son.

"We need to get rid of him at any price. You are telling the absolute truth", agreed the *nazir's* son.

They made a plan to kill the herdsboy, and they arranged an ambush in the mountains. The *nazir's* son flung a spear and killed the brave *djigit*, and the vizier's son hid the corpse in a deep cave.

For a long time the khan's youngest daughter was searching for her missing husband, but without success.

Meanwhile in a distant region, at the *djigit's* native village, his father the old khan saw that on the ceiling, where the arrow shot by his son was protruding, it began to drip blood. "Oh woe is me! A misfortune has happened to my son!" cried the khan.

His son and retainers came running at the cry. "I will ride off right now to look for my brother. Saddle my horse!" ordered the son, and he got going on his way at once.

He eventually reached the Nart's castle and had not had time to step over the threshold before the beautiful girl threw herself on his neck and shouted, "How long I have been waiting for you!"

"You are mistaking me for my brother! But where is he?" said the young man. The girl told him all that she knew. "My brother is already dead", said the young man, and he dissolved into tears.

"Do not cry, young man. I know how to help you. The wicked Nart had a ring that will return to a man his life and youth. Here it is. Take it, find your brother, and have a woman put the ring on his right little finger. And come back as quickly as you can".

The young man thanked the beauty, took the ring and rode on

further. Quite a lot of time had passed before one night he asked for a night's lodging in a certain hovel near a sheep enclosure.

"Who are you?" answered a woman's voice.

"Give shelter to a traveller, O kind woman", said the guest.

"Enter in peace", and the woman let him in.

The young man was very tired, and so he refused supper, collapsed on the bed and fell asleep. Awaking, he saw that the mistress was lying next to him. The guest silently unsheathed his sword, laid it between himself and the woman and again went fast asleep.

In the morning the mistress asked, "O husband, why aren't I pleasing to you?" He realised that this was his brother's wife, and he told her everything. "We will look for him together until we find him!" exclaimed the khan's daughter.

They were looking for a long time, and at last they found the cave where the murderers had thrown the *djigit*. The wife put the ring on his right little finger and the dead man came to life at once. And now there were many tears, and even more joy.

At home, the *djigit* arrayed himself in the red *cherkeska* and set off to the khan.

"Who are you, *djigit*?" asked the khan.

"I am the one who was favoured with the hand of your youngest daughter, and who fought with your enemies in your difficult hour".

"He is lying, this pitiful fraud. Order him to be thrown into prison!" cried the vizier's son and the *nazir's* son.

"Since when have my slaves become the khan's counsellors? All right then, show your *kleymos*!" shouted the *djigit*. Together with his brother, he threw himself on his foes, pulled up their shirts, and everybody saw the slave marks. And now the *djigit* told the khan everything.

"Put these men in chains and throw them into prison!" ordered the khan, and the vizier's son and the *nazir's* son were taken away immediately.

After that there was a big feast and a double wedding. The khan gave away his daughter in marriage, and his son-in-law's brother married the Nart's prisoner.

After the wedding both brothers and their wives set off to their native land, to their father, the khan. They say that the feast that started there is still continuing until the present time.

***

Once again, what we find in this tale, are elements from more than one tradition. For example, the old man whose wife is unable to conceive reminds us of the story of Abraham in the Old Testament. Abraham is blessed with a son, but then, as a test of his faith, he is required to offer him up as a sacrifice to God, just as the khan will be required to do.

However, it is not God who blesses the khan with an heir, two in fact, but a stranger who somehow knows what is making him sad, and offers him two apples to give to his wives. In that the encounter between the two of them takes place on a journey, the unknown man can be said to play the role of a spirit helper in this tale.

Although, contrary to popular misconception, the fruit Eve was tempted with in the Garden of Eden was not necessarily an apple (in fact, we are not given its name), there are many other associations between apples and fertility or for restoring youth. For example, "Among the Kara-Kirghiz, barren women [would] roll themselves on the ground under a solitary apple-tree in order to obtain offspring" (Frazer, 1993, p.120). And in Scandinavian mythology, Iduna, the daughter of the dwarf Svald and the wife of Bragl, was guardian of the golden apples. These were tasted by the gods whenever they wanted to renew their youth. There is also the story of Prince Ahmed in *The Arabian Nights*. The apple purchased at Samarkand in the tale (not so distant from Daghestan) provides a cure for every disorder.

One of the sons, on reaching the age of fifteen, is not to be offered to god as part of the agreement, as in the biblical tale, but to a wicked Nart. Before his departure the doomed son shoots an arrow from his bow into the ceiling, as this can be used by his brother to predict what befalls him. Shooting arrows for such a purpose can be found in many tales from the Caucasus, and is simply one of a number of methods that were employed for divination. The fact that he knows how to do this however, would indicate that he is gifted in some way, and it was also one of the techniques that shamans could use in their work on behalf of their communities.

On his way to the Nart's palace, the "doomed" son encounters a venerable old man, perhaps the very same spirit helper his father once met, who advises him on how to deal with the evil magician. The approach he recommends is for the boy to take advantage of the fact that it is a custom for all guests to be accorded the very best hospitality possible, and for them not to be asked to do any work. By following the old man's advice, the boy eventually succeeds in killing the Nart and taking his horse, which serves as the vehicle for the next stage of his journey. He also releases a beautiful girl who had been held captive by the Nart, and instructs her to remain there in the castle until his return. More about her later!

He then gets a job as a herdsboy with the khan's herdsman. The youngest of the khan's three daughter notices, and falls in love with him, She is attracted by his golden hair, which no doubt marks him out as being different and sets him apart from others. And this is what we tend to find in indigenous communities where shamanism is practised, that the prospective shaman is generally singled out in some way,

sometimes from birth, by a physically distinguishing feature such as an extra finger, a harelip or a birthmark. By the same token, the stigmata may take the form of mental or nervous dysfunction. A child who, at puberty, has fainting fits will often be regarded as a prospective shaman. So, too, will the victim of epilepsy, hysteria or the so-called "Arctic sickness", a loss of mental balance said to come about through prolonged exposure to sub-zero temperatures and boundless snowscapes (Rutherford, 1986, pp.34-35)

The three sisters then pick three apples for their father, but only the youngest daughter's apple is "ripe and tasty", indicating that she's "in the flowering of her youth". As well as being associated with fertility, apples are also found in folktales that are concerned with the preservation of youth. *The Earth will take its Own*, from neighbouring Georgia, is just such a tale (see Berman, 2007).

The time has come, the khan decides, for his daughters to marry, and the youngest daughter, against his wishes, chooses the herdsboy, and is then shunned by the rest of the family. However, after a while the khan falls ill, and only the herdsboy is able to cure him by finding the meat of the white djeyram for him. Likewise, the herdsboy, as the golden-haired horseman dressed in red, is the only person able to save the khan's army from defeat in battle.

The husbands of the youngest daughter's two sisters, not wanting to lose face, therefore plan to kidnap and kill the "herdsboy" so the khan will never find out what he did. It is at this point that the arrow the "herdsboy" shot into the ceiling in his father's home all that time ago, starts to drip blood, and his twin brother then rides to his rescue.

When he reaches the evil magician's castle, the beautiful girl mistakes him for his brother. Once she realises who he is though, she helps him by giving him the dead Nart's magical ring. He then encounters his brother's wife, who also mistakes him at first for his twin, but then helps him to find the cave where the murderers had thrown our hero. The wife puts the ring on his right little finger and the dead man is restored to life once again.

The khan, father of the three sisters, has the two evil husbands locked up, the herdsman's twin brother marries the beautiful girl who had been held captive by the Nart in his castle, and equilibrium is thus restored to the community once more.

During the course of the journey that is undertaken by our golden-haired hero in this story, not only does he face and overcome evil with aid of a spirit helper, but he is shown to have the power to heal as well by being the only person able to find the meat of the white djeyram. He also visits, and returns from the Land of the Dead with the aid of a second spirit helper in the form of the beautiful girl who had formerly been held captive by the evil magician. All this, once again, indicates that what we have here this is another example of a shamanic story rather than just a fairy tale.

Neo-shamanism (also known as "modern shamanism", "new shamanism", "urban shamanism" and "contemporary shamanism") can be said to be a mix of shamanic traditions taken from different cultures, blended into a new complex of beliefs and practices. Rowena Pattee (Kryder) PhD., a popular writer on neo-shamanism and workshop leader, defines a neo-shaman as "a modern person whose experiences of dying to the limited self and of the resultant ecstasy lead to self-empowerment and sacrifice for the benefit of his or her community" (Pattee, 1988, p.17). However, whether the techniques learnt are necessarily used to benefit the community is open to question as many take up the practices purely for personal development.

All ritual sequences can be seen to arise out of some condition of social disunity, actual or potential, and such disunity can often provide the starting point to folktales too. Within indigenous communities, however, unlike what takes place in neo-shamanic circles, ritual is "pre-eminently concerned with the health of the corporate body, with securing balance and harmony between its parts ... rather than [with] individual men and women" (Turner, 1981, p.270) and this could well result in individuals being required to subordinate their individuality to their responsibilities to the community. In other words, the shaman is primarily concerned with the modal state of the community. For example, as Hoskins points out in the case of the Kodi, the ultimate goal of the rite is to repair social relations and "heal the group" (see Hoskins, 1996, pp.287-288). In neo-shamanic circles, on the other hand, the reality is that there is no such sense of there being a corporate body of people who live and work together on a daily basis, which clearly impacts on the form neo-shamanic rituals take.

And in this particular tale, the concern of the "shaman" is not his personal development, but the equilibrium of the community. This can be seen both from the way he gives himself up to the evil magician so as to honour his father's agreement, and also from his concern for the well-being of his father-in-law even though it results in his "death".

Chapter 6

Bear's Ear

There lived or did not live a certain khan. He had a daughter, renowned for her beauty.

Every day the khan's daughter and her maidservants would go out to have a stroll in the shady palace garden.

One day, when they were enjoying themselves, a large bear jumped over the fence, seized the khan's daughter and disappeared instantly.

For a long time the bear was dragging the girl through thick forests and over dangerous cliffs. Finally they reached the cave where the bear lived.

The khan raised the alarm. His *nukyers* searched everywhere, but they were simply unable to find the missing girl.

For a long time the khan's daughter was living in the bear's cave. She could not run away, because every day the bear used to lick her heels until they bled. With time she gave birth to a boy with bears' ears, and they called him 'Bear's Ear'. He grew up quickly: a day passes, and for him it is like a month; a month is like a year. Soon Bear's Ear became a *bogatyr* [a mythical hero].

One day, when the bear had gone away hunting, the son asked, "Tell me, Mother, how did you land up in this cave?" And the khan's daughter related everything, without concealing anything.

Bear's Ear became angry and decided to take revenge on his cruel father. When the bear came back home, the son broke off a big piece of the cliff, hurled it at the bear and smashed his head. After that the son and his mother left the mountains.

At the first stopping-place Bear's Ear said, "Mother! You go to your father now and stay there. The proud khan will not accept a grandson who was sired by a bear. The earth is large, and even I will get some little corner of it". And Bear's Ear went away.

He walked by day and he walked by night and he reached a big town. He started walking through the town and shouting loudly, "Who needs a workman?"

The report reached the khan that a young man of a *bogatyr's* size and with bear's ears, was looking for work. "Why don't you bring him here", he ordered. "I want to have a look at the man with bear's ears". The *nukyers* quickly searched out the young man, and he appeared before the khan.

"Who are you, and what can you do?" asked the khan.

"I am called Bear's Ear, and you can find out very quickly what I am useful for, if you accept me for work". The khan was willing, and Bear's Ear began to work as a farm-labourer for him.

One day the khan decided to send a hundred men into the forest for firewood, and Bear's Ear said to him, "Why, khan, are you fitting out so many men for the forest? Give me food for a hundred men, and I will lay in so much firewood for you as you have never even dreamed of".

"But how will you cope with it on your own?"

"That is my business. If you are not satisfied, then there is your sword and my neck" [an expression meaning a willingness to place one's fate in another person's hands].

The khan agreed, and Bear's Ear ate up a dinner for a hundred men and set off into the forest. He tied a stout rope around a hundred plane trees, pulled them and tore them out by the roots. Then he heaped them up into a huge pile and dragged them into the town, breaking fences, sheds, and even houses.

Bear's Ear came to the palace and started shouting, "Eh, Khan! Open the gates a bit wider!" The khan looked out of the window and saw the pile of firewood as big as a whole mountain.

The khan got frightened and had the thought, "This man has extraordinary strength. Supposing he does me some harm. We need to get rid of him as soon as we can". So the khan decided to destroy Bear's Ear. "Listen, Friend!" said the khan. "Beyond that high mountain there lives a *khart*. She owes me an old debt: one *kali* [a measure of dry substances, weighing approximately 12kg] of beans".

The *khart* was very ferocious and had already killed many a *djigit*. Bear's Ear got to the lair of the *khart* and started shouting in a threatening tone, "So how about giving our khan his *kali* of beans, otherwise I will drag you to him!"

"Go into the house and collect the debt", answered the *khart*.

When Bear's Ear went into the house, the *khart* pointed to a huge chest and said, "There!"

The young man bent down over the chest, but at the same time the *khart* grabbed him by the legs in order to stuff him into the chest.

"Ach, you accursed one!" shouted Bear's Ear. "I am going to play a trick on you!" And with these words he seized the *khart*, threw her into the chest and slammed the lid. The *khart* begged him to let her go, but Bear's Ear loaded the chest on his shoulders and set out to *the palace.*

The khan saw his workman and asked, "Did you get payment of the debt from the *khart*?"

"She refused to pay it, the accursed one. But instead of the beans, I dragged her to you herself". And Bear's Ear let the *khart* out.

He khan got frightened at the sight of the monster and began shouting, "I need neither the beans nor the *khart*! For the sake of Beched [a pagan Avar god], deliver me from her".

Bear's Ear shrugged his shoulders, but he also fulfilled the khan's request; he carried the chest back, let the *khart* out, and gave her a good kick.

Meanwhile the khan was forever thinking about how he could get rid of Bear's Ear. Once he called him and ordered, "Beyond the mountain, in the forest, there lives an *azhdakha*, who owes me a young ox. Do not hang around, but go and fetch what he owes me".

Bear's Ear set off to the *azhdakha* and said to him, "Are you going to keep on cheating the khan for much longer? How about you giving him back his young ox!"

The *azhdakha* became indignant at this impudence, and he threw himself at the bold fellow.

Out of his eyes showered big sparks, from his nostrils came smoke in puffs, and from his mouth shot a hot flame. But Bear's Ear grabbed the *azhdakha* by the ears, like a kitten, and dragged him to the khan.

The khan was shaking in his shoes from fear, and he began to howl in a heart-rending voice, "I need neither the ox nor this monster! For the sake of Beched, take away the *azhdakha*".

"Damn the *kharts* and the *azhdakhas*! How much longer must I mess about with them? And what a fine khan, too! He himself does not know what he wants", muttered Bear's Ear. But he let the *azhdakha* go, and the latter galloped off into the forest like an arrow.

Now the khan summoned his viziers and began to ask their advice about how to get rid of Bear's Ear. Finally they decided to attack him with the whole army. The khan called the *bogatyr* and said, "Take the bald mare from my stable and take her off into the mountains. And what's more, you feed her up so well that she will put on weight and become round and smooth, like a hen's egg".

And so Bear's Ear led the old overworked mare into the mountains. And after him went all the foot soldiers and horse soldiers of the khan, showering the *bogatyr* with clouds of red-hot arrows. "Leave me in peace!" shouted Bear's Ear. "I am a peaceful man, and I am grazing the khan's mare here".

"Try protecting yourself and we will see what will happen to you now!" shouted the khan, hiding behind the backs of his *nukyers*.

"Ah, so you are here too, you idler!" exclaimed Bear's Ear.

"Well I am going to teach you a lesson!" Without stopping to think, he tore the khan's mare into four pieces. Two of them he threw at the army, and immediately devastated half of it. With the other two he started to destroy those who were left alive. The army could not withstand the onslaught, and they took to their heels, with the khan in front of them. It is said that they are still running, and they will soon be at the end of the world.

Meanwhile, Bear's Ear set off for distant regions. He walked by day, and he walked by night, and one day he met a strong hefty fellow who was carrying two huge plane trees on each shoulder.

"What strength!" exclaimed Bear's Ear.

"What kind of strength do I have! Now Bear's Ear, who dragged the *khart*-giantess to the khan, they say has real strength", answered the big fellow.

"Bear's Ear, that's me".

"Then I am your fellow-traveller!" exclaimed the big fellow joyfully, and the two of them began to stride on together.

They are walking and they see a mountain man sitting on the road and easily revolving a huge millstone on his knees. "What a hero!" exclaimed the friends.

"Am I really a hero? Now Bear's Ear who, so they say, brought the *azhdakha* to the khan as if he was a kitten, is really a hero", answered the mountain man.

"And here he is, Bear's Ear", said the big fellow with the plane trees.

"Then allow me to be your fellow-traveller", asked the mountain man.

After long wanderings they found a small forest where there were plenty of animals to hunt and clean water, and they stayed and lived there.

On one occasion Bear's Ear and the millstone turner set off hunting. Their third companion was sitting by the fire and boiling meat in a huge cauldron. Suddenly a lame hare came running up to the fire, and sitting and riding on the hare was some kind of short, tubby bearded fellow. He was the size of a *lokot* [a measure, now no longer in use, equal to one cubit or the length of the forearm-about 50cm] and his beard was the length of three *lokots.*

"Salam aleikum! How about you treating me to some meat".

"Vaaleikum assalam, help yourself".

The tubby fellow ate up an enormous piece of meat and demanded more. When he was refused, the tubby fellow pulled a hair from his beard and tied up the strong hefty fellow really tightly. Then he ate all the meat, climbed upon his hare and galloped away.

Bear's Ear and his companion came back home tired and hungry. But instead of supper they heard the sad news of what had happened.

On the next day the rotator of the millstone stayed home, while Bear's Ear and his companion set off hunting in the forest. This time too, everything happened as on the day before. The bearded man approached on his hare, easily tied up the strong man, ate up all the meat and galloped away.

On the third day Bear's Ear stayed home. He lit the fire and began to boil the meat. When it was nearly cooked, the short tubby fellow rode up and demanded, "How about treating me to some meat!"

"Certainly not!" answered Bear's Ear.

The tubby fellow tore a hair from his beard and rushed at Bear's Ear, but the latter grabbed the tubby one and tied him up by the beard to a large plane tree.

When his comrades returned home, Bear's Ear wanted to show them the prisoner. But how could he! The bearded fellow had already torn out the plane tree by its roots and run away.

"To the chase!" shouted Bear's Ear, and all three rushed to catch up with the tubby fellow. For a long time they were wandering through the forest thickets, and at last they arrived at a deep hole.

"He is in the hole. Look, next to it is lying that same plane tree!" exclaimed Bear's Ear, and he asked for ropes to be brought. His comrades tied the ropes round him and began to lower him into the hole. The hole was very deep. At first, it began to blow deathly cold from the hole, and then it blazed with heat, but Bear's Ear withstood both the one and the other. At the bottom there turned out to be a vast palace of gold and silver, and in one of the rooms Bear's Ear saw a girl of unprecedented beauty. She was radiating such a bright light that the room was light, like in the day. Right there the tubby fellow was sleeping.

"Who are you? Run quickly, before my husband wakes up and kills you!" shouted the girl.

"Beched alone is in charge of life and death!" exclaimed Bear's Ear, and he grabbed the tubby fellow by the beard. The bearded one woke up and threw himself at the newcomer with a loud cry. Bear's Ear tore off the tubby fellow's beard, and he banged him against the wall and killed him.

"Tell me, my beauty, who are you and how did you land up in this dungeon", said he, turning to the girl.

"I am the daughter of a powerful *padishakh* [an oriental king]. But this magician kidnapped me and forced me to be his wife", answered the beauty. "My father will reward you generously for freeing me".

"What is your father's treasure to me? Rather than all of that, I would prefer your attentions", answered Bear's Ear.

"O steadfast young man! My attentions are not altogether a worthy reward for your bravery. There will be nobody in my heart besides you. Lead me where you will".

They collected up the bearded fellow's treasure in a huge sack and tied it to the rope. "Haul it up, friends!" shouted Bear's Ear. And his friends pulled the sack out of the hole.

"And now it is your turn", said Bear's Ear to the girl.

"No, it is better that you go up first", she begged him. "I have a presentiment of something bad".

For a long time they were arguing, but finally the girl agreed to go up first. "If anything happens, then you should know what to do", she warned. "Into the palace will come two rams, a black one and a white one. The white one will carry you out up above, but the black one will drag you away down to the underground kingdom. And do not forget to take the magician's sword".

The beauty tied the ropes round herself and was pulled up to the top. "Is there anything else left?" the comrades shouted from above.

"Only me!" shouted Bear's Ear. "All pull together". But the rope was slowly slipping down to the bottom, and he realised that he had been betrayed. Bear's Ear began to rush about in rage, but there was nothing he could do. Soon he lay down and went to sleep.

Towards morning two large rams entered the room. Bear's Ear was very glad and jumped on to one of them; but in his haste, instead of the white one he sat on the black one. With a crash Bear's Ear fell through into the underground kingdom and he fell on the roof of a poor old woman's house. For a long time he was lying on the roof, but at last he got up and saw a vast fine-looking town.

"Ey, kind woman! Give me a little water", he asked the house-owner.

"What's up with you? Have you fallen down from the earth?" she answered. "Aren't you ashamed to hurt an old woman's feelings? What kind of discussion can one have about water?!"

"Is there really no water in the underground kingdom?" said Bear's Ear in surprise.

"We have as much water as we like. But a huge nine-headed *azhdakha* has settled at the spring, and only lets us get to the water once a year. In return for that, he is given each time the most beautiful girl".

"To hell with that! Give me, mother, two large pitchers", asked

Bear's Ear. "I will get some water for you".

"Come to your senses, little son, do not go to certain death. After all, many of those whom the *azhdakha* has killed were not at all inferior to you in strength".

"All right then. If it is so fated, then I will perish, but if not, I will stay alive", said Bear's Ear. He took the pitchers and set off to the reservoir.

By the water was lying a terrible monster. Bear's Ear approached, and began to fill up the pitchers. The *azhdakha* did not say a word; he only stared straight at the brave young man. On the next day Bear's Ear again went for water, and the *azhdakha* again stayed silent.

At that time word of the dare-devil reached the khan of the underground kingdom. He summoned Bear's Ear to the palace and asked him to free the underground kingdom from the monster. "You are our only hope. As a reward you will receive everything you wish", said the khan.

"All right!" answered Bear's Ear. "I am prepared to join battle with the *azhdakha*, provided you make me some felt ears".

In the morning Bear's Ear buckled on the sabre of the bearded magician, fastened on the felt ears, took the pitchers and went to the spring.

This time the *azhdakha* said, "O son of a man! The first time I let you pass, as a guest. The second time as a friend. But you unashamedly turn up here a third time too. Is it possible that you do not value your life?"

"And your days are numbered too, you damned *azhdakha*! You are the one who has no shame: you have deprived the people of the water that Beched gave them, and you swallow girls alive! Beware, you creature!"

The *azhdakha* struck first, but only the brave fellow's felt ears fell off. Bear's Ear swung the sorcerer's sabre, and all nine heads of the *azhdakha* rolled on the ground. He took the monster's heads and carried them to the khan.

In the underground kingdom there began unheard of celebrations. Some wept, others laughed, yet others broke into a dance. And then they all rushed to the water and greedily began to drink. "How can we reward you, my son?" asked the khan of the underground kingdom. "If you want to, you can marry my daughter and be my successor. If you want to, take all my treasure".

"Thank you for your kindness", answered the hero. "To become related to you would be a high honour for me. But the treasure, I do not need. Just help me to get myself back up to the earth".

The khan lowered his eyes and said quietly, "I am not able, my son, to fulfil your request. But in our forest there lives an old eagle. Perhaps he will be willing to help you". And the khan sent some of his people to the eagle, but soon they came back with his refusal.

Then Bear's Ear set off to the eagle. On a massive plane tree he found the nest of the large eagle. Going nearer, he suddenly saw a black snake crawling up the tree. Bear's Ear killed it and stood waiting for the eagle.

Suddenly a terrible wind arose and, like a cloud, the eagle landed on the plane tree. He saw the slain snake and exclaimed, "Son of a human, you have killed my most terrible enemy, who has destroyed our offspring every year. Ask for what you want, and any request you make will be fulfilled".

"In that case, take me back to the earth", asked Bear's Ear.

"All right", answered the eagle. "Slaughter fifty buffaloes, stock up with water, and then I will fly you where you want".

Fifty of the largest buffaloes from the khan's herd were slaughtered, and from their skins were sewn *burdyuks* [vessels made from whole goat or ox skins; used for the storage and transport of wine or vodka] and they were filled with water. One of the eagle's wings was loaded with the meat, and the other with the water. Bear's Ear took his leave of the underground kingdom, and started off on his way.

They were flying for a long time, and Bear's Ear gave the eagle the meat to eat and the water to drink. They had already almost reached the earth, when the exhausted eagle asked for meat for the last time. Bear's Ear saw that all the provisions had run out, so he cut off a big piece from his leg and gave it to the eagle. The eagle swallowed down this meat, made one final effort and reached the earth.

Bear's Ear climbed down from the eagle, and the eagle saw that he was lame. "Why are you lame, friend?" asked the eagle. Bear's Ear then told him everything. "That is no problem", answered the eagle. He spat out the piece of meat, spat on it a few times and placed it on the strong man's leg. The leg became whole at once.

Bear's Ear thanked the eagle and began to search for the place where his comrades had betrayed him. Soon he heard shouting and swearing and saw that his former friends, like cocks, were fighting over the girl. At the same time she was sitting and continually repeating that she needed nobody except her rescuer.

Bear's Ear quickly put an end to this quarrel. Under his blows one of the traitors fell face downwards, and the other fell on his back. It is said that they are lying like that to this day.

Meanwhile Bear's Ear and his bride went to the khan her father, who received them with joy. A magnificent wedding was arranged, and there were many guests from all over the world. And Bear's Ear is alive even now. He sits alongside the khan and performs many glorious deeds.

***

Totemistic and magical ideas and beliefs have an ancient tradition in Daghestan, and are reflected in the folktales from the region.

Many ... examples could be given of encounters with bears which are presented as fact, as events occurring in reality: thus in the village of Apshi, Buynaksk region, a bear abducted a woman, and when a child was born to them, the bear came to the village and carried a cradle off to the forest [from field notes made by P. Magomedova, a student at the Daghestan State University]; in Tsunta region a bear escaped from an avalanche and forced his way into a mill, where the miller offered hospitality to him in the most friendly way for a whole week [from diary notes made by Kh. Karimova, a correspondent of the newspaper "Red Banner"]; in the same region a bear snatched away by a woman working in a garden in broad daylight and demanded that she cohabit with him, but did not cause her any physical harm [from field notes made in the village of Bezhta by the ethnographer M. Dibirov]; a bear fell in love with a woman from Urchukh, Sovetskiy region, and came to the village every night to see her [from field notes by the ethnographer M. Dibirov] (Abdurakhmanov, 1992, pp.400-401).

Evidence that behind all these stories lie the most ancient totemistic ideas and beliefs

is provided above all by cultic concepts with magical functions, like keeping a bear's skin at home (for protection against illness), in a barn (for the wellbeing of the cattle) or in a cemetery (to keep cattle away), giving the bear skin to someone to wear during the performance of the rain-making ceremonies, and so on (Abdurakhmanov, 1992, p.401).

The potency of bears did not go unnoticed either, and was believed to be transferable to humans:

In the village of Almak, Kazbekovski rayon, old men say that formerly childless parents pierced the fangs of a bear and hung them by threads across a cradle, and when they had a child, placed them under his pillow. Later the fangs would be passed on to another such couple (Chenciner, 1997, p.193).

As for the burning of hair, it

> is one of the most archaic forms of divination, widely practised in world mythology. Current among the peoples of Daghestan are a whole series of contradictory beliefs connected with hair: hair and nails may not be cut at night, or burned on the fire, they must be buried in the ground or hidden in a wall; if someone secretly burns a tuft of hair together with a scrap of clothing they may see their future spouse in a dream, though this is considered sinful and liable to punishment (Abdurakhmanov, 1992, p.403)

The suggestion that Bear's Ear is gifted with special powers is reinforced on the journey by the two men he encounters along the way. Although the first is able to carry two huge plane trees on each shoulder and the second to rotate a huge millstone on his knees, both point out that their abilities are nothing compared to those of Bear's Ear, while being unaware of the fact that it is Bear's Ear they are actually talking to at that point in time.

The same theme can be found in a Chechen variant of this story–*Chaitong, the Son of a Bear*, which can be found in *Tales of the Ingush and Chechens*. Chaitong meets a young man who, with his little finger, was playing with a huge log", and another who "was sitting on the ground and listening in to the conversation of the ants". Both, however, say that "The real marvel is what Chaitong, the bear's son, accomplishes" The two men become the companions of Bear's Ear and travel with him until they come to "a deep hole" which provides a means of gaining access the lower world. There they help Bear's Ear by lowering him into the hole, where he kills the bearded fellow and, at the same time, rescues the *padishakh's* daughter whom had been kidnapped by him.

We learn that when Bear's Ear is lowered into the hole, "At first, it began to blow deathly cold from the hole, and then it blazed with heat, but Bear's Ear withstood both the one and the other". And this gives us further evidence as to his special powers.

Excessive heating of the body can be seen as a means of assimilating sacred power and the ability to withstand extreme heat is one of the characteristic marks of shamans. "In whatever cultural context it appears, the syndrome of magical heat proclaims that the profane human condition has been abolished and that one shares in a transcendent mode of being, that of the Gods" (Eliade, 2003, p.72).

In fact, throughout history, spiritual practitioners have resorted to the metaphor of heat or fire to describe the process of transformation that leads to enlightenment, and there are many examples of stories about heat-immunity feats in Stith Thompson's Motif-Index of Folk-Literature. For example, D1841.3, "Burning magically evaded" and D1841.3.2.3, "red hot iron carried with bare hands without harm to saint" (see McClenon, 2002, p.71).

As for the figure of the shaman, he / she can be said to have pursued submission to the world of the spirit, often combined with severe ascetism in the same way as was adopted by the holy fools in Christianity and in Sufism. In Japan, for example, shamans can only obtain the special powers they need to bridge the gap between the two worlds through ascetic practices known in general as *gyo*–fasting, cold water treatments, and the recitation of words of power (see Blacker, 1999, p.85). The demonstrations that are organised by yamabushi to convince the community that the discipline has risen above the ordinary human state include *hi-watari* or fire-walking, *yudate* or pouring boiling water over the body, and more rarely *katana-watari* or climbing up a ladder of swords. The ladder of swords symbolizes the ladder to heaven, and only the shaman can mount it.

On our hero's attempt to return to the middle world, he ignores the padishakh's daughter's warning not to trust the two companions waiting for him in the world above, and the result is that they help the girls to ascend but leave Bear's Ear behind.

It is at this point in the story that we learn of the girl's special powers. Not only is she gifted with second sight, but she is also a lot less naïve and more worldly than Bear's Ear: "If anything happens, then you should know what to do", she warned. "Into the palace will come two rams, a black one and a white one. The white one will carry you out up above, but the black one will drag you away down to the underground kingdom. And do not forget to take the magician's sword".

Unfortunately, in his haste, instead of the white one Bear's Ear sits on the black one. All is not lost however, for as a reward for overcoming a huge nine-headed *azhdakha* (a monster that usually lies in wait by a river or a spring, only allowing people to take water if, in exchange, he is offered a girl as a sacrifice), the local khan agrees to help him, and he does so by directing our hero to an old eagle. Bear's Ear kills the eagle's most terrible enemy the snake, and in return the eagle then agrees to carry him back to his own world. In the process, however, when the eagle runs out of food on the journey, he has to feed her with his own flesh.

In the tales of the Vainakhs (the collective local name of the Ingush and Chechens) the idea of three worlds is met quite often. The hero of a tale and of epic poetry lands up in the lower world and there continues to accomplish the heroic deeds that he accomplishes in the sunlit (upper) world. In the underground world the positive hero meets three sisters (daughters of a prince or *padchakh*) kidnapped by an evil monster; he fulfils their mission and frees them from misfortune. The role of the evil monster is performed by a dragon (*sarmak*), and that of the hero's [spirit] helper is [played by] the eagle (*erzi*). (Malsagov, 2007, p.260).

The story of *Bear's Ear*, apart from there being only one kidnapped daughter needing to be rescued, and the fact that the *sarmak* is replaced by an *azhdakha*, is more or less identical to the tales that Malsagov refers to above, and it can be seen they are clearly related to each other.

It has been suggested that

> The process of dismemberment culminating in consumption is a plain survival from the hunting phase, since it is exactly what the hunter does to his quarry; namely kills it, skins it, removes the offal, then cooks and eats it ... By offering himself as a sacrificial victim the shaman is repaying the debt humans have incurred by their slaughter of animals. (Rutherford, 1986, pp.36 & 37).

However,

> As with every other detail of the shaman's calling, the death/rebirth cycle, though constant, shows regional variations so that in some places it is reduced to the symbolical form of an initiatory rite. In others, the extreme opposite applies and the candidate may have to undergo ordeals of an extremely painful nature, either self-inflicted or inflicted by initiators (Rutherford, 1986, p.38).

In this particular story, the ordeal entails the prospective shaman feeding the eagle with parts of his own body.

As for the conclusion to the tale, on his return to this reality, Bear's Ear disposes of the two traitors who were supposed to be his friends, and then goes to the khan with his bride who receives them with joy and arranges a magnificent wedding for them. In this way the equilibrium of the community is restored and the shaman's journey is thus complete.

Chapter 7

Arsuman

There was once a couple who had one son and one daughter. The son was called Arsuman.

One day the wife was ill and said to her husband, "I would like some meat." "What kind of meat would you like?" he asked. "I would like some of Arsuman's flesh," she answered. The man took his son, killed him and gave his wife to eat. The daughter said when she came home, "Mother, I am hungry." "There in the corner is your soup, take it and eat," answered the mother. But as the daughter ate her soup, she saw a little finger lying in it. "That is my brother's little finger," she said, rolled it in a napkin and carried it to the church. There the little finger changed into a bird. The bird flew away, and went first to a draper. "What will you give me if I sing you a little song?" asked the bird. "A piece of silk," answered the merchant. And the bird began to sing:

> I am a little bird, bird, bird,
> I am a Tirilili, tirili!
> My father killed me,
> My mother ate me,
> My little sister let me fly away.
> I am a little bird.

When he had got his silk he flew away to a shop where needles were sold and he asked: "What will you give me if I sing you a song?" "A packet of needles," was the answer. And again the bird sang:

> I am a little bird, bird, bird,

I am a Tirilili, tirili!
My father killed me,
My mother ate me,
My little sister let me fly away.
I am a little bird.

When he had got his needles, he flew to a shoemaker, sang his song and got a pair of shoes. Then he flew to a pinshop, sang his little song and got a paper of pins. From there he flew to the roof of his father's house, perched on it and called: "Father, look up here!" The father said: "But perhaps you will take your revenge on me! I am frightened." "Don't be frightened," answered the bird. "Hold a sieve in front of your face and look up."

And as the father looked up, the bird threw the needles in his face and blinded him. Then he called his mother and blinded her with the pins. But then he called his little sister and told her to hold up the hem of her dress, he wanted to give her something. And he threw the silk into her dress and then the shoes, and flew away and was never seen any more.

*****

Arsuman is an Udian story, taken from Dirr, A. (1925) Caucasian Folk-tales, London: J.M. Dent & Sons Ltd. (translated into English by Lucy Menzies). The Udians are described in the Introduction to the volume as being "a small people with Lesghian tongue. They only inhabit two villages, Warthaschen and Nisch, east of Nuchi in Trans-Caucasia".

Though mention is made of the sister of Arsuman carrying her brother's little finger to the church, presumably for some form of healing, the origin of the tale could well date back to much earlier times, and be based on an account of a shamanic initiation, in which the initiate is consumed and then transformed, ready to take on a new role within the community. The fact that Arsuman shape-shifts into a bird is significant too, as this would enable him to fly between worlds and thus act as an intermediary.

"Etymologically, 'religion' comes from the Latin word '*religio*' (*religere*, which literally means 'to tie together again,' i.e., to reunite the creation with the creator)" (Heinze, 1991, p.137). Acting in his / her role as an intermediary, this is what the shaman can be said to do. And it was as an intermediary for others that they traditionally made their living.

Ordeals the apprentice may be required to undergo in non-ordinary reality during his / her initiation can include torture and violent dismemberment of the body, the scraping away of the flesh until the body is reduced to a skeleton, the substitution of viscera and renewal of the blood, and even a period spent in Hell during which the future shaman is taught by the souls of dead shamans and by demons (see Eliade, 2003, p.96).

The process of dismemberment culminating in consumption can be seen as a survival from the hunting phase, since it mirrors what a hunter does to his quarry:

> By offering himself as a sacrificial victim the shaman is repaying the debt humans have incurred by their slaughter of animals. In the words of Ramusen's Eskimo, "All the creatures we have to kill and eat, all those we have to strike down and destroy to make clothes for ourselves, have souls, as we have" (Rutherford, 1986, p.37).

Nostalgia for an initiatory renewal can be regarded as "the modern formulation of man's eternal longing to find a positive meaning in death, to accept death as a transition rite to a higher mode of being … not subject to the destroying actions of Time" (Eliade, 2003, p.136). Indeed, it can be argued that it is only in initiation that death is given a positive value. And the practitioners of neo-shamanism, by attempting to make the techniques accessible to everyone rather than just an elite few, would claim that they thus enable us to experience such a process for ourselves.

Pagan cults of the death and resurrection of the god of fertility were common to agrarian peoples worldwide. With the Kabardians, Cherkess and Adyghes in particular, the *azhegafa*, or ram, played an important role in the festival of the First Furrow. With his band, he went around all the homes of the village, performing his 'death and resurrection' for a reward or ransom paid by the owner of the house ... The resurrection of the ram symbolised growth of grain from freshly seeded once-abandoned soil (Chenciner, 1997, p.190).

The "pattern of sacramental death and rebirth is universal. It occurs not only among shamanistic hunting societies but also among the planting societies, where it is incorporated into the cycle of the crops, and the dying god is equivalent to the ear of corn that falls into the ground to die, and is born again in the new growth" (Larsen, 1998, p.63). In the Babylonian myth of creation, the vegetation that dries up and withers during the dry season is interpreted as Tammuz' descent to the underworld, and when rain and fertility return to the land Tammuz is resurrected and the cosmos is created anew (see Otzen, Gottlieb, and Jeppesen, 1980, p.15). The same motif is evident in the Ba'al myth in Ugarit, in which Mot represents the ripened grain, and Anat deals with him accordingly so the cycle continues. It can also be found recorded in our own tradition, in the ballad of *John Barleycorn* (see Berman, 2008b, pp.95-96).

The Caucasus is a land with which the earliest folklore of Europe is connected. One only has to think of the Argonauts and of Prometheus to be reminded of the long ages through which the mountainous country between Pontus and the Caspian Sea, between Europe and Asia, has been connected with man's inherent love of storytelling (Dirr, 1925, p.v).

The stories that have presented in this study offer ample evidence to show that this is the case.

Chapter 8

Kitschüw

When the Russians occupied the Caucasus in the early nineteenth century, the Czar sent a representative there whose job was to report back on the ethnic groups to be found in the region. The official observed that the Jews living in the mountains were unlike Russian Jews in their outward appearance, had different lifestyles, wore different clothes, and even spoke a different language. To distinguish them from their Russian counterparts, he referred to them as "Mountain Jews" in his report, and this explains the origin of the nickname (see Mikdash-Shamailov, 2002, p.11).

According to legend, the *Juhur* as they called themselves, are descended from the ten lost tribes that were exiled from the Kingdom of Israel in the eighth century BC. Although this cannot be verified, we do know there were Jewish communities in the eastern Caucasus as early as the third century AD. They moved there to escape persecution in Persia and brought their own language with them. (see Mikdash-Shamailov, 2002, p.17).

The language ... is an ethnolect of Tat, which belongs to the southwestern subgroup of the Iranian language. Four local dialects of the ethnolect can be identified: those of Derbent, Kuba, Makhachkala-Nalchik (historically the Haytaghi dialect), and Vartashen – now Oguz (historically, the Shirvani dialect). The usual designation for this ethnolect in the scientific literature is "Judeo-Tat," but the Mountain Jews call it *zuhun juhuri* – literally, "the Jewish language." (Mikdash-Shamailov, 2002, p.37).

Although they are these days relatively few in number (3,600 in total, according to http://russia.rin.ru/guides_e/4753.html [accessed 6/8/08]), during the Middle Ages the Derbent Region of Daghestan was known as *Cufut-Dagh*, or "Mountain of the Jews", which suggests that not only were they indigenous to this region but also that they must have been numerous at one time (see Blady, 2000, p.159). Small communities of Mountain Jews can also be found in both neighbouring Chechnya and Azerbaijan.

The possible origins of the Mountain Jews are expanded upon in the following quote:

> Armenian and Georgian chronicles report the first Jewish movements into the Transcaucasus at the beginning of the 6th century B.C.E. Some of the first arrivals were probably captives sent as gifts to friendly rulers by Nebuchadnezzar of Babylon. Later, there were Jewish refugees fleeing into the Caucasus-presumably through Mesopotamia and Persia-after the destruction of the Second Temple.

> Before the 1917 revolution the Mountain Jews belonged to the most conservative Jewish group ... In olden times, however, mountain Jewish scholars, according to a rabbinic tradition, participated in the creation of the Talmud. There is a mention of a Rabbi Nahum Hamadai (the name itself indicates that he came from Mydia, later Azerbaidzhan) and of Simon Saphro from Derbent.

> In contrast to other Jewish groups, there are no priests or Levites among the Mountain Jews, and most of the names of both men and women date back to the epoch of the wanderings of Israel in the Arabian Desert or to the period of the Judges and Kings. In this fact, [some would say a confirmation can be seen] ... of the Mountain Jews' own strong oral

tradition of descent from the lost ten tribes of Israel (Halle, 1946, p.545).

Although the Mountain Jews only form a small community these days, at one time Judaism was the predominant religion in the region. The choice of Judaism as the state religion in pagan Khazaria can be explained by the presence in the country of a large local Khazar-Jewish population, of Jewish proselytes among the mountaineers and the Khazars, and by the desire of the Khazar khans themselves to show that they were politically independent of hostile neighbouring states, of the Muslim Arab caliphate, and of Christian Byzantium. Another important factor in the acceptance of Judaism by the Khazar khans was the influence of the Jewish aristocracy: merchants, magnates, and rabbis serving at the courts of the Khazar khans as businessmen and advisers. This led to the emergence of a Jewish-Khazar kinship entity.

After the fall of the Khazar Khanate to the Arabs from the south and the Russians from the north toward the end of the tenth century, many Khazar Jews withdrew into the depths of mountainous Daghestan; those who remained in their old haunts found themselves in an oppressive feudal dependency on the Arab rulers of the Caucasus and their local agents. They were forced to bring tribute and other payments, and, to preserve their Jewish faith, to pay a special tax (*j'dzh*); many of them, particularly the converts from among the mountaineers and the Khazars, turned to Islam. The Arab caliphate and their agents were in turn replaced by new conquerors (the Seljuk Turks, the Persian shahs, and Turkish sultans) and a series of Azerbaijani and Daghestani khanates and overlords. In conditions of feudal disintegration, the Mountain Jews found themselves under the control of local rulers with the legal status of dependent peasants. With the unification of Azerbaijan and Daghestan with Russia in 1813, the Mountain Jews accepted Russian citizenship, the status of "Jew" was imposed on them, and they began to be called into military service.

The social oppression of czarism, to which were added the pogroms, then made life extremely difficult for the Mountain Jews. To compound their problems even further, the White Guard bands of Bicherakov and Denikin, invaded the area in 1918-1920, and were responsible for the destruction and looting of a series of Mountain Jewish villages. This led to many of the families migrating to Palestine (adapted from http://www.everyculture.com/Russia-Eurasia-China/Mountain-Jews-History-and-Cultural-Relations.html [accessed 1/8/08]).

As with all ethnic groups in Daghestan, village life for the Mountain Jews revolved around the clan and the village commune.

Because of the alleged longevity of the people and early marriages, a clan could easily encompass four or even five generations, reaching seventy-five to 100 in number. The oldest male was the head of the family. When he died, the leadership was passed on to his oldest son. Each nuclear family consisted of a man and his wife, or wives, and children. If a man had more than one wife, each wife and her children occupied a separate hut. Petty conflicts were usually ironed out by the arch-patriarch of the clan. If he could not resolve a matter, the case would be brought before the *nesi'a haeadah* (community chieftains) (Blady, 2000, p.163).

The authority of the head of the family was absolute, and all the important decisions, such as arranged marriages, purchases, who to reward and who to punish, the division of property, and even the allocation of the salaries earned by the sons of the family, were taken by him alone. This continued well into the 20th century, and only started to break down as a result of the economic, social, and political changes that were brought about by the October Revolution (see Mikdash-Shamailov, 2002, p.123).

To redress the balance to a certain extent, it should perhaps be pointed out that "Although the father (*bebe*) was, from time immemorial, the absolute ruler of the extended family (*kele-kiflet*), on Fridays his authority was assumed by the mother (*dedey*) or grandmother (*kele-dedey*), who then reigned supreme over the household" (Mikdash-Shamailov, 2002, p.79).

One of the roles of fathers was to teach their sons combat and how to use weapons as soon as they were old enough to defend themselves, and like other fighting peoples of the region,

Mountain Jewish men dressed in the classic "Circassian" style. This consisted of an outer garment called a *tcherkeska*, which was made form homespun cloth. Silver-tipped cartridges and holders were sewn on both sides across the breast of the *tcherkeska*. A Mountain Jewish man also wore a *beshmet* (long white shirt buttoned up high, with a round stand-up collar). The leather belt was tight and supported the armature, which consisted of a powder horn, a pistol, and a saber. The trousers were narrow, and the boots had no heels. A black *papak* (sheepskin cap) rounded out his attire. Weapons were worn even when praying in the synagogue (Blady, 2000, pp.165-166).

On the subject of hospitality, a national institution in the Caucasus,

As one of the oldest inhabitants in the region and the people who brought monotheism to the Caucasian soil, it may well have been the Jews who wove the biblical patriarch Abraham's practice of *hachnosat orchim* (welcoming guests) into the fabric of Daghestani culture. Every guest was treated as if he were personally sent by God. In every Jewish home a special room or hut covered with the finest carpets was set aside for guests.

Every host would protect his guests with his life, and lavish on them the finest food and spirits. The host and his family might forego eating meat in order to provide the guests with this luxury (Blady, 2000, pp.165-166).

Indeed, so great was the importance of what is known as *ginog* or hospitality "that not even the poor could forego the possibility of housing guests in a separate room" (Mikdash-Shamailov, 2002, p.126).

As for the form of Judaism practised by the Mountain Jews, they have incorporated many pagan rites and superstitions into their belief system. For example, in the cycle of wedding, birth, and funeral rituals are a number of pre-Judaic and pre-monotheistic concepts, including belief in the purifying strength of fire, water, amulets, and talismans against evil spirits (water nymphs, devils, etc.).

Num-Negear is the lord of travellers and home life, Ile-Nove is Ilya the Prophet, Ozhdegoe-Mar is the house spirit, and Ser and Shegadu are malicious spirits. And to honour Gudur-boy and Kesen-boy (the spirits of autumn and spring) the Mountain Jews arranged special festivals. The Shev-Idor celebration is dedicated to Idor, master of the plant kingdom, and for the eve of spring festival girls would practise a form of divination that made use of flowers taken from forests (adapted from http://russia.rin.ru/guides_e/4753.html [accessed 6/8/08]).

The majority of Mountain Jews today are non-believers, and one of the reasons for the departure from the faith was the increasingly negative attitude in the former Soviet Union as a whole to the Jewish religion, partially in reaction to the creation of the state of Israel. The more conservative elements in the community began to link the leading elements of the Mountain Jewish population with Zionists, and this damaged the Jewish ethnic identity (constitutionally the equal of other ethnic groups). This also explains why many Mountain

Jews began not only to conceal their Jewish faith but to call themselves "Tat." Many of them, even believers, stopped attending the three synagogues in Daghestan (in Derbent, Makhachkala, and Buynaksk), and they are now only used by a small number of believers, primarily of the older generation. As a result, the faith is mainly maintained through the performance of traditional rituals in the home (adapted from http://www.everyculture.com/Russia-Eurasia-China/Mountain-Jews-Religion.html [accessed 1/8/08]).

As for religious holidays, it would probably be true to say that are observed more because of tradition than belief. Many traditional funeral and memorial customs are still practised, though not in the same way as they are by Jews in other countries. For example, during the *shiva* period (seven days of mourning after death) it is said that

> the soul refuses to believe that it has just been separated from the body, and therefore still hovers over the house in which the person has just died. Acting on this belief, the Mountain Jews would light a candle and place a chair on the spot where the deathbed had been located. The candle was placed on a copper tray that in turn was placed on a stool. The tray symbolized the partition between the divine in heaven and man on earth (Blady, 2000, p.167).

Kitschüw is a Tatian story, taken from Dirr, A. (1925) Caucasian Folk-tales, London: J.M. Dent & Sons Ltd. (translated into English by Lucy Menzies). Tatian is described in the Introduction to the volume as being a modern Iranian dialect, and is spoken by the so-called mountain Jews in Daghestan, from one of whom this story was collected. Folktales (*ovosune*), like the example presented below, were related by professional storytellers known as *ovosunechi*.

***

There was one king in Schura[3], another in Petrowsk. He in Schura had five sons, but he of Petrowsk had none. Both kings were great friends; they were always to be seen together. One day they were both in Schura at the wedding of a nobleman. Suddenly a messenger arrived from Petrowsk with the tidings that the queen was ill and begged her husband to return home. The king pulled a hundred roubles out of his pocket, gave them to the messenger, and in half an hour he and his friend were already in Petrowsk. But they had not been long there before a messenger came post haste from Schura to say that the Queen of Schura was laid up, and would like her husband to return home at once. This messenger too received a hundred roubles reward and the two kings travelled together back to Schura. There they vowed to each other that if one of them should become the father of a son and the other the father of a daughter, the two children should marry each other and should be formally betrothed as son as they were born.

And it so happened that the Queen of Schura bore her sixth son, the Queen of Petrowsk her sixth daughter. The children both grew very fast; when the princess was still quite a little girl she already understood about looking in a mirror, and the prince could ride at the same age. When the princess was thirteen years old she was so beautiful that the queen would not let her be seen on the street, and at that age the prince had already begun to hunt.

Now what more shall I tell you? Shall I tell you of the three-headed Ajdaha?[4]

Well, this Ajdaha had heard how beautiful the princess was, and so he asked his mother, who could work magic, how he should set about winning her for his bride. The mother changed him into a little golden bird, and said, "Fly away to the roof of the king's palace." This he did, and just as he flew up the princess was standing at her window. The golden bird flew to her, and the princess, who had never seen a bird like him, caught him at once. But the bird, who had quickly changed himself back into an Ajdaha, seized her and carried her off.

Now of whom shall I tell you next? Of the queen, the mother of the stolen princess?

Well, the queen came soon afterwards into her daughter's room and saw that it was empty. She had the whole town searched at once, but in vain ... the princess was nowhere to be found. Messengers were sent to Schura, but she was not there either. The five eldest sons of the King of Schura set out to look for her, and as they passed through a wood they saw their youngest brother, Kitschüw, lying under a tree. "Hullo, brother!" they called out to him, "why are you lying there? Do you not know what a misfortune has happened to you? Your bride has disappeared!" "Ah, ha!" replied the brother, "but I know better than you who has carried off Altuntschatschä (Golden Locks). It was the three-headed Ajdaha." The brothers proposed to him when they heard this that they should all go together to rescue the princess. But first of all they would go home to make their preparations.

Both kings happened to be in Schura at the time. "Hullo, father! Hullo, father-in-law, listen to me." Called the youngest son; "what are you going to do, father-in-law, to find Altuntschatschä again?" "I will give you seven Tulpan horses," he answered. "And you, father, what do you think of doing?" "I will give you weapons and provender for your journey."

"And you, oldest brother, what will you do?" "I will pray to God to divide the sea that we can see in what part of it the Ajdaha is." "And you, second brother, what will you do?" "I will pray to God to build a tower in which we can hide ourselves if the Div follows us." "And I," said Kitschüw himself, "I will cut off Ajdaha's head."

The following day Kitschüw set out, with his five brothers and his friend Alsanä. Mounted on the Tulpan horses provided by Kitschüw's father-in-law, they travelled in seven days a distance that would on other steeds have taken them seven years, and at last they came to the shore of a great sea. "Now, eldest brother, fulfil your promise!" said Kitschüw. He did so, and at his prayer the sea was opened up and they saw the place where the three-headed Ajdaha was. He lay at the bottom of the sea asleep and Altuntschatschä had one of his heads on her lap. Kitschüw drew his sabre and rushed at Ajdaha to slay him. "Wait a little, Kitschüw," said Altuntschatschä, "you cannot do it that way. Do you see that fish there? Kill it first: you will find in it a box which contains Ajdaha's soul. Take it out and throw it away; then he will not be able to get up, and you can cut off his heads and set me free." Kitschüw followed the advice of his bride, caught the fish, opened it up, pulled out the box, cut the soul in pieces, struck off Ajdaha's three heads, took Altuntschatschä and fled.

What shall I tell you now? About Enäj, who was Ajdaha's mother?

She came some time later to visit her son. When she found him cut to pieces she was very angry and began to kill the fish of the sea. Then a big fish swam up to her and said, "Enäj, why are you so angry at us? Look there! That big fish ate up the soul of your son, kill him and take your son's soul out of him!" Enäj did as she was advised, and by that means brought her son back to life again. He wept at first when he became alive again, but then made ready at once to follow Kitschüw.

"Brothers!" cried Kitschüw. "Brothers, do you know where this fine rain comes from all of a sudden? I know. Ajdaha is coming after us. Now it is your turn, second brother, to keep your word." And the second brother prayed to God, and God at once let a high tower appear before them, in which the six brothers and Altuntschatschä hid themselves. When Ajdaha came to the tower he sprang up it but could not reach them. He sprang a second time, and Kitschüw succeeded in cutting off all his three heads with one blow.

Seven days later Kitschüw with his bride and his seven brothers arrived home. How glad the kings were and what a joyful feast they arranged!

Kitschüw lay in bed for a time and Altuntschatschä sat beside him. Suddenly he sprang up and said, "I must go away at once." "Where must you go, then, so suddenly?" asked his bride. "Wait for me for three years, three days, three hours and three minutes," he replied: "if I am not back by that time, then you can do whatever you like." He said the same thing to his mother, who wept bitterly when she heard her son's intention. Aslanä wanted to go with him, but Kitschüw would not allow him, and set out quite alone on his journey. On the ninth or the tenth day he came to a place where three paths divided, and there were three signposts. On one of them was written, *If you go by me, you will not return*; on the second, *If you go by me, you may not return*, and on the third, *If you go by me, you will return*. Kitschüw chose the first path.

After a time he came to a spring. There his horse said suddenly in human speech, "Kitschüw, dismount and bathe in this spring." "Why?" asked Kitschüw. "Why?" replied the horse, "you do not need to know that just now." Kitschüw got off and bathed in the spring. After a time the horse began again to speak: "Kitschüw, I will tell you something. In the country in which we now are there are two kings who are brothers. The elder brother is very fierce, the younger not so much so. I will change you into a golden bird. The two brothers will soon pass by this way. I will disappear now, but I will give you three hairs out of my tail and tell you two words which you must remember carefully. If you say the first, you will change yourself into sand; if you say the second, you will change into a grain of corn. And if any great danger threatens you, set fire to one of the hairs." So saying the horse disappeared, and Kitschüw set himself in the form of a bird at the edge of the spring.

Shortly after, he saw two riders coming; they were the two kings. The younger of the two was a keen sportsman; wherever he saw a beautiful bird, he either killed it or caught it in order to eat it. But this time his brother warned him: "Leave that bird alone! Do not catch it, or do it any harm." But the golden bird flew around the young king, who said to himself: "How can I pass such a lovely bird without catching it?" So he put out his hand and the bird flew straight into it. The huntsman at once put the little creature in his breast, without his elder brother noticing it. When they got home, he gave the bird to one of his sisters. In the evening she was playing with the bird when it flew to her shoulder, pecked her first on the cheek and then on the breast. "No, no," she said, "my breast is not for you, that belongs to Kitschüw." Hardly had she spoken before the bird took on human form; the same Kitschüw of whom she had just spoken stood before her. "I am Kitschüw," he said. "Good," she answered, "but before I can belong to you, you must accomplish three things.

First you must wrestle with me, then you must change yourself into sand, and then you must fight with seventy Ajdahas." Kitschüw forced the maiden to the ground, then he changed himself into sand, but before he entered into the fight with the seventy Ajdahas he set fire to one of the hairs from the tail of his horse. It appeared at once, and together they overcame the seventy and slew them all. Then he mounted his horse, put the maiden before him and rode home, where a great feast was arranged to celebrate his return.

But when he wanted to go to his wives-he had two wives now-he was not allowed, he must first fetch the sister of the three Ajdahas for his friend. He set out at once. First he killed the three Ajdahas. But their sister sat in the middle of a great tree. He must first of all tear it open. That he did, but afterwards was very tired and laid himself down to sleep. In the meantime his friend came along. He took the tree with him and rode off. But Kitschüw wakened at once and hastened after him. When he saw it was his friend, they went happily homewards together.

And when they arrived a much greater feast was held than had ever been held before.

***

Interestingly, even though it was told by a Jewish informant, the story displays virtually no adaptation to the Jewish oikotype (cultural environment) within Daghestan. One oikotype that does appear is that the two kings pay a messenger in roubles, which is an adaptation to the oikotype of Russian-dominated lands. But the Jewish oikotype is not there. However, this is understandable in that a story about chiefs ("kings") suited the non-Jewish environment and, at the same time, Jewish listeners could comprehend it and even be gratified by it without any adaptation to Jewish culture being required. (The notion of oikotype was introduced by Carl Wilhelm von Sydow, 1948, in 'Geography and Folk Tale Oicotypes'. In *Selected Papers in Folklore*, Copenhagen: Rosenkilde and Bagger, pp. 44-59).

The fact that at the start of the tale the two kings agree their two children should marry each other and that they should be formally betrothed as soon as they were born was nothing out of the ordinary.

Traditionally all marriages were arranged. Most families employed a professional matchmaker (*ilche*) a mature woman respected by the community. Matchmaking was sometimes performed by the families themselves at a much earlier stage, before or shortly after their children were born. Occasionally, a girl was betrothed and conveyed to her future husband's family at the age of seven or eight, several years shy of the permissible age for marriage–twelve years and a day. She would assist her future mother-in-law in running the household and become accustomed to her new family (Mikdash-Shamailov, 2002, p.95).

The style employed by the storyteller is what has been described as "magic realism", and what happens at the start of the tale is nothing particularly out of the ordinary until the Ajdaha is introduced. The ability of the Ajdaha to shape-shift into a bird and then back into its original form, and the soul theft of the princess alert us however as to what we can expect.

Worthy of note is the manner in which the storyteller engages, and actively involves, the listener by introducing questions: "Now of whom shall I tell you next? Of the queen, the mother of the stolen princess?" – "Yes, please do!" being the likely response.

It is also interesting to note that the five eldest sons find their youngest brother lying under a tree, thus more than likely having been in a sleeping (i.e. in a trancelike state), and already knowing what has befallen the princess. In other words, by seeming to have the power of divination, he is already marked out as being different from the others, as was traditionally the case with the "initiate" or shaman-to-be.

Why seven horses, rather than six or eight? And not only are there seven horses for Kitschüw, his five brothers, and his friend Alsanä but they travel for seven days too. The number has been chosen for a reason:

Seven is a mystic or sacred number in many different traditions. Among the Babylonians and Egyptians, there believed to be seven planets, and the alchemists recognized seven planets too. In the Old Testament there are seven days in creation, and for the Hebrews every seventh year was Sabbatical too. There are seven virtues, seven sins, seven ages in the life of man, seven wonders of the world, and the number seven repeatedly occurs in the Apocalypse as well. The Muslims talk of there being seven heavens, with the seventh being formed of divine light that is beyond the power of words to describe, and the Kabbalists also believe there are seven heavens–each arising above the other, with the seventh being the abode of God.

The number seven was also of great importance to the Mountain Jews: In the synagogues on *Hoshana Rabbah*, congregants bearing Torah scrolls make seven circuits around the bimah (the platform where the scrolls are placed to be read from), on *Simhat Torah* a procession headed by the rabbis makes seven circuits round the bimah, seven benedictions are recited as part of the marriage ceremony, at the funeral of a religious functionary the procession stops seven times, two candles lit at burial ceremonies are left to burn for seven days, and there are seven ritual days of mourning.

Although the cosmology, described in Creation Myths, will vary from culture to culture, the structure of the whole cosmos is frequently symbolized by the number seven too,

> which is made up of the four directions, the centre, the zenith in heaven, and the nadir in the underworld. The essential axes of this structure are the four cardinal points and a central vertical axis passing through their point of intersection that connects the Upper World, the Middle World and the Lower World. The names by which the central vertical axis that connects the three worlds is referred to include the world pole, the tree of life, the sacred mountain, the central house pole, and Jacob's ladder (Berman, 2007, p.45).

Even at this early relatively early point in the plot, the indications are that what we have here is essentially a shamanic story rather than what at first sight might appear to be just a simple fairy tale, and the same can be shown to be the case with many other tales from the region see Berman, 2007, for two examples from neighbouring Georgia.

It was the belief in various indigenous shamanic communities that death to the physical body was the result of the departure of the soul, and this is also no doubt why Altuntschatschä advises Kitschüw to deal with Ajdaha's soul before cutting off his three heads. For without his soul he posed no danger.

On shamanic journeys however, everything is possible as the rules of ordinary reality no longer apply, which explains how Enäj is able to bring Ajdaha back to life again. Kitschüw then turns to his second brother for assistance, who conjures up a tower, a common feature of the landscape in the region. Chenciner even includes a photo of one, a 16th century Chechen single family blood feud siege tower at Tazbichi, near Itumkale, on the road to Daghestan, which was presented to him as a gift (see Chenciner, 1997, p.149). From the vantage point of the tower, Kitschüw is able to put a permanent end to Ajdaha this time, and the number seven appears again, as this is the length of time it takes him to return from there with his bride.

While in bed, perhaps through a dream and thus in a trancelike state, Kitschüw gets a calling which results in him leaving Altuntschatschä. On the subsequent journey, he opts for the most dangerous of the three paths available to him, displaying the kind of courage one would expect of a shaman. Due to the eristic nature of such practices, death was always a strong possibility in such an occupation. His horse, in human speech, then directs him as to what he should do next. This ability to understand the language of the animals and to communicate with them is a traditional attribute of the shaman, and marks him out once again as being different from the other brothers. Kitschüw is told by him to bathe in a spring, considered to be a place of power in the Caucasus, in order to cleanse himself in preparation for the "sacred space" he is about to enter. The horse, acting as Kitschüw's spirit helper also gives him the ability to shape-shift, again an ability traditionally associated with shamans.

Water plays an important part in a number of Jewish rituals.

On Rash ha-Shanah, for example, (which, for the majority of Jews, marks the beginning of the New Year) the Mountain Jews observe the *tashlikh* ceremony. This involves people saying prayers near a source of water so as to "cast" their sins into it. There is also an old Caucasian Jewish practice that entails expelling bad dreams by "telling them to the river" (see Mikdash-Shamailov, 2002, p.81).

Not only water, but fire is highly regarded by the Mountain Jews as a benevolent power too. "Their views about fire may have been influenced by the Persian Zoroastrian religion. A Jew who wishes to impart additional force to an oath will say, "I swear by this fire" (Mikdash-Shamailov, 2002, p.117) It is also forbidden to throw certain objects into a fire such as hair, onion peel or bread, as it is believed this might cause it to become contaminated..

By shape-shifting into a golden bird, Kitschüw meets the woman destined to become his second wife. First, he has to succeed in the tasks she sets him, which he does with the aid of his spirit helper, his horse. Not only that, but Kitschüw also finds a wife for his friend Aslanä, thus restoring equilibrium to the community, which is of course the main function of the shaman.

In the tale we find references to a number of different religious beliefs and practices, and given the fact that the folktale is the product of a group of Iranian speaking Jews living in a Moslem country where pagan practices are still prevalent, this is hardly surprising. These include reference to Judaism, with the oldest son praying to God (thus clearly a monotheistic one) to divide the sea to assist his youngest brother, just as he once did for Moses.

Then there is the fact that Kitschüw is allowed to have two wives, which is of course permissible under Islamic law. In practice, however, in Daghestan monogamy has been usual for a long time now, mainly "because the ratio of men to women only varies up to 100/105, providing too few local women" (Chenciner, 1997, p.54).

Although now not acceptable for Jews to have two wives, and it is more commonly known as an Islamic practice, Lamech (see Genesis 4:19) is the first of a long line of biblical men who did have more than one, and it seems that God approved of such marriages: Abraham had two wives at the same time, Sarah and Hagar (see 16:3), as well as several concubines (see 25:6); Esau "takes" two wives in 26:34, and then another in 28:9; Jacob had two wives and two concubines (see 32:22), and Esau (Isaac's son) had several wives too (see 36:2).

As for the Mountain Jews, there is actually a tradition of polygamy among them: "Before the 1920s, the community of Mountain Jews did not accept the ban on polygamy enacted in the Middle Ages by Gershom ben Yehudah. Bigamy was practised by wealthy people and rabbis, even if their first wives were not barren" (Mikdash-Shamailov, 2002, p.103). Each wife would occupy a separate house or, less often, a separate room in a house shared by the whole family. In fact, by all accounts, it would seem that remaining single was hardly an option for a man: "The notion that a man should remain unmarried was anathema to the Mountain Jews. Such a person was not considered to be a full-fledged member of the community. He was forbidden to carry weapons (Blady, 2000, p.165). Therefore, even within a Judaic community the practice of having more than one wife is not unheard of, and can help to explain why Kitschüw has two in this tale.

References can also be found to pagan beliefs and practices, with a magical cause being assumed for the coming of rain:

"Brothers!" cried Kitschüw. "Brothers, do you know where this fine rain comes from all of a sudden? I know. Ajdaha is coming after us.

Offerings to invoke rain or the sun have long been made in the Caucasus, and rituals to bring about rain are still conducted in Daghestan today:

In Karabudakent, after a prolonged drought, a meeting was announced from the minaret. Those with milk herds were asked to bring milk, and the rest to bring produce for the sacrificial meal. In a separate ceremony the women went up to their sacred place, while the men sacrificed a horned animal at the sacred well, dressed in skin coats turned inside out. They chanted a special prayer, shouting "Yes, we will have rain!" and then poured the water from the well over each other. Young men were made to jump into the cold water, after which they all went to the mosque for a meal.

Until the 1900s, all over Daghestan, the mullah presented his village with a sheep's shoulder bone with Arabic inscriptions to encourage rain. In Khurik, the inhabitants gathered in their holy place, and after the sacrifical meal, they ran to the river and poured water over each other. The Lak had a variety of scapegoat totems: straw-doll, straw-cat or straw-donkey. The Avar poet, Rasul Gamzatov wrote that when the earth cracked in the heat, trees drooped, and fields dried up and plants, birds, sheep and people longed together for heavenly water (his Tsada) villagers picked a boy as a rain-donkey. Dressed up in coloured cloths, like grass faded by the sun, he was led on a rope through the village by children, chanting to Allah for rain. The women of the village ran out after the rain donkey, with jugs or basins and poured water over him, while the children responded "Amen, amen!" (Chenciner, 1997, p.105).

The Mountain Jews would perform what is known as the *gudil* ceremony, which has two functions: to appeal for rain during a drought or, conversely, to bring excessive rains to an end.

Until recently, children in some regions marched in a noisy procession carrying a *gudil* figure–a scarecrow made of branches and rags. The procession wound its way past all the Jewish homes, where young women laughed and older women recited prayers. All distributed presents to the children and poured cold water on the scarecrow, hoping the rain would similarly quench the land (see Mikdash-Shamailov, 2002, p.117)

Sometimes the child was replaced by an adult member of the community dressed as a scarecrow, and when the appeal was for sunshine, the scarecrow was coloured red.

The mix of traditions found within the same tale very much reflects the lifestyles of the people. Although "The traditional religion of the Mountain Jews is Judaism, ... a number of pre-Judaic and pre-monotheistic concepts, including belief in the purifying strength of fire, water, amulets, and talismans against evil spirits (water nymphs, devils, etc.)" are still practised. As for oaths, they "are rendered by the Torah and the Talmud, but also by the hearth," which would be the traditional Daghestani way (taken from http://www.everyculture.com/Russia-Eurasia-China/Mountain-Jews-Religion.html [accessed 1/8/08]).

This helps to explain how, though they are basically monotheistic, the Mountain Jews also believe in the existence of supernatural forces, but under God's dominion: "These forces may be visible, and some take on the form of animals. They have the power to punish people for their sins and reward them for their good deeds. While some have permanent features, others change in appearance and character according to a person's behaviour (Mikdash-Shamailov, 2002, p.112).

These supernatural beings include *She'atu*–"an evil spirit that stirs in the dark and often appears in deserted houses, near water sources, in open fields, or in the thick of forests" (Mikdash-Shamailov, 2002, p.112). There is also *Num-ne-gir* (which translates as "she whose name shall not be uttered") who "is associated with fertility and family happiness. She is also responsible for the safety of travellers. Her name remains unspoken because it inspires great fear" (Mikdash-Shamailov, 2002, pp.112-113). Then there is the spirit of the Water Mother, known as *Serovi* or *Dedei-ovi*, who "lures lovers or pregnant women to a river in which they disappear forever" (Mikdash-Shamailov, 2002, p.113).

Belief in the evil eye is also widespread, with people attempting to counter its effect by carrying amulets prepared by rabbis or healers, or verses copied from the Talmud.

Healers are thought to have the ability to understand and cure physical weaknesses, a condition usually attributed to the evil eye. One method they use is to throw balls of dough into a fire while calling out the names of all enemies known to the patient. If the balls burn without exploding or bouncing out of the fire, the evil eye is not to blame. If a ball explodes when the name of an enemy is called out, the healer must rip off a piece of the enemy's garment, burn it, let he patient inhale the smoke, and sprinkle the patient's forehead with the ash. The evil eye can also be thwarted by reading a glass of water. Wax or melted lead is tossed into the water, and the ensuing form is believed to closely resemble the person who caused the illness. Salt is then scattered over a fire, blinding–or even destroying–the enemy whose image is revealed (Mikdash-Shamailov, 2002, p.111).

Notwithstanding such supernatural beliefs, at the same time the Mountain Jews honour all the festivals and holy days that are recognized by Jews worldwide. Additionally, they also celebrate a Spring Festival, called Shegme-Vasal (the Spring Candle) in Daghestan, which is unique to the Caucasian Jews.

Fire plays a special role in this Festival. Young people in Daghestan customarily skip over a campfire and chant songs of supplication, asking to be protected from the evil eye, their fields shielded from drought during the planting season, and their animals spared from the cattle plague. The campfire is lit on the site of an abandoned threshing floor to ensure that the coming year will be fertile" (Mikdash-Shamailov, 2002, pp.85-86)

So despite the well-documented attempts by the Soviet regime to erase the distinctive features of its member nations and homogenize them into a single Russian model, the Mountain Jews have still managed to maintain many of their traditional customs in addition to adopting a number of the local ones, and the story of *Kitschüw* reflects all this.

Bibliography

Abdurakhmanov, A. M. 1992. "Totemistic Elements in Rituals and Legends about Animals." Hewitt, George (ed.). *Caucasian Perspectives*, pp. 392-405. Lincom Europa, Munich.

Ausubel, N. (ed.) (1948) *A Treasury of Jewish Folklore*, London: Vallentine Mitchell.

Beard, M. (1993) *Mythos in mythenloser Gesellschaft: Das aradigma Roms*, ed. Fritz Graf (Stuttgart and Leizig: B.G.Teubner, pp.62-64).

Berg, H. (2001) *Dargi Folktales: Oral stories from the Caucasus and an introduction to Dargi grammar*, Universiteit Leiden, The Netherlands: Research School of Asian, African, and Amerindian Studies.

Berman, M. (2006) 'The Nature of Shamanism and the Shamanic Journey', unpublished M.Phil Thesis, University of Wales, Lampeter.

Berman, M. (2007) *The Nature of Shamanism and the Shamanic Story*, Newcastle: Cambridge Scholars Publishing.

Berman, M. (2008a) Soul Loss and the Shamanic Story, Newcastle: Cambridge Scholars Publishing.

Berman, M. (2008b) *Divination and the Shamanic Story*, Newcastle: Cambridge Scholars Publishing.

Berman, M. (2008c) *Tell Us A Story*, Folkestone: Brain Friendly Publications www.brainfriendly.co.uk .

Blacker, C. (1999) *The Catalpa Bow A Study in Shamanic Practices in Japan*, Japan Library, an imprint of Curzon Press Ltd.

Blady, K. (2000) *Jewish Communities in Exotic Places*, Northvale, New Jersey: Jason Aronson Inc.

Booker, C. (2004) *The Seven Basic Plots*: Why we tell Stories, London: Continuum.

Calinescu, M. (1978) 'The Disguises of Miracle: Notes on Mircea Eliade's Fiction.' In Bryan Rennie (ed.) (2006) *Mircea Eliade: A*

Critical Reader, London: Equinox Publishing Ltd.

Chenciner, R. (1997) *Daghestan: Tradition & Survival,* Richmond, Surrey: Curzon Press.

Chenciner, R, Ismailov, G., & Magomedkhanov, M., (2006) *Tattooed Mountain Women and Spoon Boxes of Daghestan,* London: Bennett & Bloom.

Colarusso, J. (1997) 'Peoples of the Caucasus'. In Encyclopaedia of Cultures and Daily Life, Pepper Pike, Ohio: Eastword Publications (taken from www.circassianworld.com [accessed 15/7/08]).

Curtis, W.E., (1911) *Around the Black Sea,* London: Hodder & Stoughton.

Czaplicka, M.A. (2007) *Shamanism in Siberia* [Excerpts from] *Aboriginal Siberia: A Study in Social Anthropology,* Charleston, SC: Biblio Bazaar (original copyright: 1914, Oxford).

Diachenko, V. (1994) 'The Horse in Yakut Shamanism' In Seaman, G., & Day, J.S. *Ancient Traditions: Shamanism in Central Asia and the Americas,* Boulder, Colorado: University Press of Colorado.

Diakonova, V.P. (1994) 'Shamans in Traditional Tuvinian Society' In Seaman, G., & Day, J.S. *Ancient Traditions: Shamanism in Central Asia and the Americas,* Boulder, Colorado: University Press of Colorado.

Dirr, A. (1925) *Caucasian Folk-tales,* London: J.M. Dent & Sons Ltd.

Dolidze, N.I. (1999) *Georgian Folktales,* Tbilisi, Georgia: Mirani Publishing House

Dow, J. (1986) "Universal aspects of symbolic healing: A theoretical synthesis", *American Anthropologist* 88: 56-69.

Eliade, M. (1964) *Myth and Reality,* London: George Allen & Unwin

Eliade, M. (1981) *Tales of the Sacred and the Supernatural,* Philadelphia: The Westminster Press.

Eliade, M. (1989) *Shamanism: Archaic techniques of ecstasy,*

London: Arkana (first published in the USA by Pantheon Books 1964).

Eliade, M. (1991) *Images and Symbols*, New Jersey: Princeton University Press (The original edition is copyright Librarie Gallimard 1952).

Eliade, M. (2003) *Rites and Symbols of Initiation*, Putnam, Connecticut: Spring

Frazer, J. (1993) *The Golden Bough*, Ware, Hertfordshire: Wordsworth Editions Ltd (first published in 1922).

Gagan, J.M. (1998) *Journeying: where shamanism and psychology meet*, Santa Fe, NM: Rio Chama Publications.

Hallam, M. "The Matrix". In *New View*, Summer 2002, London: The Anthroposophical Association Ltd.

Halle, F.W. (1946) 'The Caucasian Mountain Jews'. *Commentary*, October 1946. Published by the American Jewish Committee, New York. Reprinted in Ausubel, N. (ed.) *A Treasury of Jewish Folklore*, London: Vallentine Mitchell.

Harner, M. (1990 3rd Edition) *The Way of the Shaman*, Harper & Row (first published by Harper & Row in 1980).

Heinze, R.I. (1991) *Shamans of the 2oth Century*, New York: Irvington Publishers, Inc.

Helman, C. (2006) *Suburban Shaman tales from medicine's frontline*, London: Hammersmith Press.

Hoskins, J. (1996) 'From Diagnosis to Performance: Medical Practice and the Politics of Exchange in Kodi, West Sumba.' In Carol Laderman & Marina Roseman (eds.) *The Performance of Healing*, London: Routledge. Pp.271-290.

Hultkrantz, A. (1992) *Shamanic Healing and Ritual Drama*, New York: Crossroad.

Hultkrantz, A. (1993) 'The Shamans in Myths and Tales', in SHAMAN Vol. 1 No. 2. Pp.39-55.

Hunt, D.G. (2007) *Avar folk tales*, compiled, translated from Avar and with commentary by D.M. Ataev; translated from the Russian with new introduction by D.G. Hunt. Russian edition published by Nauka, Moscow, 1971 as part of their series: "Tales and myths of Oriental peoples" (in loose leaf format in the British Library).

Ingerman, S. (1993) *Welcome Home: Following Your Soul's Journey Home*, New York: Harper Collins Publishers.

Jung, C.G. (1968 2nd Edition) *Psychology and Alchemy*, London: Routledge & Keegan Paul.

Jung, C.G. (1977) *The Symbolic Life*, London and Henley: Routledge & Keegan Paul.

Kremer, J.W. (1988) 'Shamanic Tales as Ways of Personal Empowerment.' In Gary Doore (ed.) *Shaman's Path: Healing, Personal Growth and Empowerment*, Boston, Massachusetts: Shambhala Publications. Pp.189-199.

Larsen, S. (1998) *The Shaman's Doorway*, Rochester, Vermont: Inner Traditions.

Lévi-Strauss, C. (1968) *Structural Anthropology*, Harmondsworth: Penguin.

Lyle, E. (2007) 'Narrative Form and the Structure of Myth' in *Folklore 33* 59.

Malsagov, A.O. (2007) *Tales of the Ingush and Chechens*, (Russian edition published by Nauka, Moscow, 1983 as part of their series: "Tales and myths of Oriental peoples").

McClenon, J. (2002) *Wondrous Healing: Shamanism, Human Evolution, and the Origins of Religion*, Illinois: Northern Illinois Press / Dekalb.

Mikdash-Shamailov, L. (ed.) (2002) *Mountain Jews: Customs and Daily Life in the Caucasus*, Jerusalem: The Israel Museum.

Pattee, R. (1988) 'Ecstasy and Sacrifice.' In Gary Doore (ed.) *Shaman's Path: Healing, Personal Growth and Empowerment*, Boston, Massachusetts: Shambhala Publications. Pp.17–31.

Rennie, B. S. (1996) *Reconstructing Eliade: making sense of religion*, Albany: State University of New York Press.

Ripinsky-Naxon, M. (1993) *The Nature of Shamanism*, Albany: State University of New York Press.

Rutherford, W (1986) *Shamanism: The Foundations of Magic,* Wellingborough, Northamptonshire: The Aquarian Press.

Segal, R.A. (2004) *Myth: A Very Short Introduction,* Oxford: Oxford University Press.

Seligman, M.E.P. (1975) *Helplessness: On Depression, development, and death,* San Francisco: Freeman

Smith, S. (2006) *Allah's Mountains: The Battle for Chechnya,* London: Tauris Parke Paperbacks.

Taube, E. (1984) 'South Siberian and Central Asian Hero Tales and Shamanistic Rituals.' In Mihály Hoppál (ed.) *Shamanism in Eurasia,* Part 1. Göttingen, Edition Herodot. Pp.344-352.

Turner, H.W. (1971) *Living Tribal Religions,* London: Ward Lock Educational.

Turner, V. (1981) *Drums of Affliction,* London: Hutchinson University Library for Africa (first published in 1968).

Turner, V. (1982) *From Ritual to Theatre: The Human Seriousness of Play,* New York: PAJ Publications (A division of Performing Arts Journal, Inc.).

Turner, V. (1985) *On the Edge of the Bush: Anthropology as Experience.* Tucson, AZ: University of Arizona Press.

Turner, V. (1995) *The Ritual Process: Structure and Anti-Structure,* Chicago, Illinois: Aldine Publishing Company (first published in 1969).

Ussher, J. (1865) *A Journey from London to Persepolis,* London: Hurst & Blackett.

Van Gennep, A. (1977) *The Rites of Passage,* London: Routledge and Keegan Paul (original work published in 1909).Vitebsky, P. (2001) *The Shaman,* London: Duncan Baird (first published in Great Britain in 1995 by Macmillan Reference Books).

Vinogradov, A. (2002) 'The Role and Development of Shamanistic Discourse among Southern Siberian Ethnic groups in the post-Soviet period,' in *The Anthropology of East Europe Review,* vol.20 No.2.

Walsh, R. N. (1990) *The Spirit of Shamanism,* London: Mandala.

Winkelman, M. (2000) *Shamanism: The Neural Ecology of Consciousness and Healing*, Westport, Connecticut: Bergin & Garvey.

Winterbourne, A. (2007) *When The Norns Have Spoken: Time and Fate in Germanic Paganism*, Wales: Superscript.

Index

FOOTNOTES

1 The word *Daghestan* or *Daghistan* means "country of mountains", derived from the Turkic word **dağ** meaning mountain and Persian suffix **-stan** meaning "land of". Most of the Republic is in fact mountainous, with the Greater Caucasus Mountains covering the south and the highest point being the Bazardyuzi peak at 4,466 m.

2 Mircea Eliade has been much maligned by academics in the field of Religious Studies in recent years. However, as Åke Hultkrantz has pointed out, he can be credited with being largely responsible for opening "wider vistas when, ignoring the common restriction of shamanism to Siberia and the Arctic regions, he included the Americas, the Indo-Europeans, South-eastern Asia and Oceania in his discussions of the phenomena of shamanism". Not only that, but Eliade's opus has also been responsible for inspiring "an intensified study of shamanic ideology and a fresh dedication to further research in the field" (Hultkrantz, 1993, p.40). Indeed, without his trailblazing work, many of us would never have got interested in the subject in the first place. Consequently, I offer no excuses for quoting from his work in this volume.

3 Temir-Chan-Schura, in Daghestan, is forty-six kilometres from Petrowsk, on the shore of the Caspian Sea.

4 A fabulous creature, something like a Div

B O O K S

O is a symbol of the world, of oneness and unity. In different cultures it also means the "eye," symbolizing knowledge and insight. We aim to publish books that are accessible, constructive and that challenge accepted opinion, both that of academia and the "moral majority."

Our books are available in all good English language bookstores worldwide. If you don't see the book on the shelves ask the bookstore to order it for you, quoting the ISBN number and title. Alternatively you can order online (all major online retail sites carry our titles) or contact the distributor in the relevant country, listed on the copyright page.

See our website **www.o-books.net** for a full list of over 500 titles, growing by 100 a year.

And tune in to myspiritradio.com for our book review radio show, hosted by June-Elleni Laine, where you can listen to the authors discussing their books.

myspiritRadio